HOW THE WORLD WORKS
ASTRONOMY

THE STORY OF
ASTRONOMY

*From plotting the stars
to pulsars and black holes*

Anne Rooney

SIRIUS

SIRIUS

This edition published in 2017 by Sirius Publishing, a division
of Arcturus Publishing Limited, 26/27 Bickels Yard,
151–153 Bermondsey Street, London SE1 3HA

ISBN: 978-1-78428-450-3
AD003879UK

Printed in Malaysia

CONTENTS

REACHING FOR THE STARS

'The history of astronomy is a history of receding horizons.'

Edwin Hubble, astrophysicist, 1936

Just 500 years ago, most people in the Western world believed that the Earth sat at the centre of the universe with everything else revolving around it. They thought humans were created to be masters of this universe and that the heavens were unchanging for all eternity.

We now know that we are an evolved and evolving species, one among millions of plant and animal species living on a planet orbiting a fairly small star in an outpost of an unremarkable galaxy, somewhere in an unknowably vast universe. We know that there are countless billions of other stars and probably billions of other planets and that the history of the universe stretches billions of years into the past. Paradoxically, with greater knowledge has come greater recognition of the limits of our knowledge. We can account for and explain only a tiny proportion of all there is in the universe. We don't even know whether there is just one universe, or many universes.

The night sky reveals the stars that have captivated the human imagination for millennia.

The story of astronomy is one of emerging knowledge and emerging ignorance. It tells how we have come to know so much about the universe and our home in it, but shows that there is much more still to discover. It is a story that has barely begun, as we stand on the brink of space exploration.

From superstition to science

Our early ancestors attempted to explain what they saw in the heavens, often using mythology alongside their careful observations and measurements. With settled civilizations came written records and mathematics, allowing more detailed observations to be made and maintained over many years. Then, around 2,500 years ago, the Ancient Greeks started to explain the cosmos without recourse to mythology or the supernatural and so began the science of astronomy.

But the separation of astronomy from the supernatural did not come at a stroke. Astronomy only gradually moved from being the province of priests to the pursuit of scientists. For centuries, the observations and calculations of astronomers were directed towards religious and superstitious ends. They were used to fix the times for prayers and religious festivals, to predict conditions and events on Earth in the political or personal spheres, and to seek propitious times to implement plans. Astrology and astronomy remained inseparable for millennia. Even in the 16th and 17th centuries, respectable astronomers often had a foot in the astrology camp. While they didn't all believe that there was any validity in astrology, they found it could be lucrative nonetheless.

The constellation Taurus, from a German astronomical globe of the 1530s.

The great divide

And then, over a period of only one hundred years, starting in 1543, astronomy and our astronomical knowledge changed beyond measure. First, two giant supernovas (exploding stars) appeared within 32 years of each other (in 1572 and 1604); none has been seen since. They demonstrated conclusively that the cosmos is not fixed and unchanging for all eternity. The old dogma had to shift to accommodate this development. Second, the invention of the telescope came just four years after the second supernova. It revealed there is far more in the night sky than we can see with our eyes alone. These events provided the vital evidence needed for a new theory of the universe to gain credence, one in which the sky is not fixed for all eternity and in which the Earth is not central. With the telescope to extend astronomers' vision, the path was clear for the development of modern astronomy.

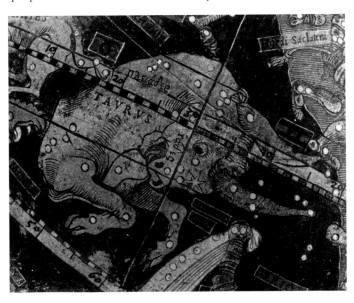

CHAPTER 1

The first
ASTRONOMERS

*'Astronomy compels the soul to look upwards
and leads up from this world to another.'*

Plato,
Greek philosopher,
4th–5th century BC

Imagine living in the Stone Age and
looking up at the night sky. What
would you notice on a clear night?
First, the Moon: a bright, shining body
that changes shape over the course of
around 29 days from new crescent to
full circle and back again, and which
moves across the sky during the night.
Next, there are a lot of bright pinpricks
of light. Without light pollution, far, far
more stars would be visible than we can
see today. You would also see a fuzzy
band of dim light that stretches across
the sky – the Milky Way.

*The Moon has shone above the Earth with the reflected light of the
Sun for four-and-a-half billion years.*

From seeing to observing

There would not be much to do at night in the Palaeolithic or Neolithic eras, so you might take to observing these objects in the sky with some care, night after night. You might then notice that most of the points of light twinkle, while a few cast a steady light. Those that twinkle move together, rotating during the course of a night around a set point. That point is not directly overhead unless you are standing at the North or South pole. You might notice that the points of light nearest the horizon rise or set over the course of the night and disappear for months of the year, reappearing predictably the following year.

You would probably notice that only a few of these points of light move along their own paths relative to the majority. Most twinkle and stay in fixed positions in relation to one another; they are the stars, originally known as the 'fixed stars'. Those that move independently and shine steadily are the planets. Long ago they were called 'wandering stars' as they seemed to wander among the fixed stars – indeed, the word 'planet' comes from the Greek *planētēs*, meaning 'wanderer'. As a Stone Age observer, you would notice how they differ from the twinkling fixed stars, but you would not be able to tell that they are fundamentally different bodies.

You might sometimes see a bright light that streams briefly across the sky and disappears – a shooting star or meteor. And occasionally you might notice, if you had been observing carefully in the past, a new star that moves across the sky slowly, night by night, before eventually disappearing. With its dim 'tail' of light trailing behind (or, actually, sometimes in front of it), this is a comet – but it's a rare occurrence.

Time lapse photography shows how the stars revolve around the celestial pole over the course of a night.

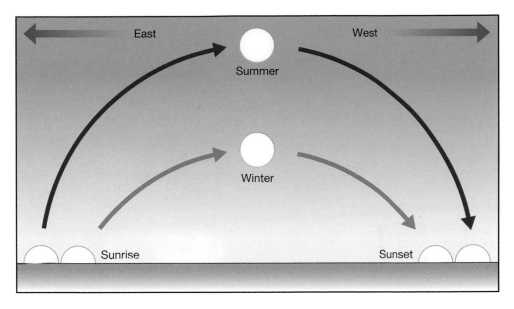

East West

Summer

Winter

Sunrise Sunset

Except at the equator, the Sun rises higher in the sky during summer than during winter.

In the daytime, the sky is dominated by just one body. You would see the Sun rise and follow a predictable path across the sky before setting at a point opposite its rising. Unless you were at the equator, you would notice that the day is longer in the summer than the winter, and the Sun travels higher in the sky in the summer.

Space and time

It would not take long for a Stone Age observer to notice that the appearance and

THE CHANGING SKY

We think of the night sky as pretty much the same every night, yet a Palaeolithic star-gazer would not see quite the same stars as we do. The northern Pole Star would not be Polaris; instead, the brighter star, Vega, would have been close to the North celestial pole (see page 12). However, as the Pole Star changes, following a cycle of about 26,000 years (see box, page 18), some Palaeolithic observers would have seen Polaris as the Pole Star last time round. Some constellations now seen only in the southern hemisphere would have been visible in the north during some months of the year and vice versa. And as all the stars are constantly moving, some would be in very slightly different places in relation to one another. This is called the 'proper motion' of the stars (see page 180), and results from each star moving on its own trajectory independent of all but those closest to it in space. But as these changes happen over thousands of years, much would appear the same to observers on Earth as it does now.

The celestial poles are found by drawing an imaginary line through the Earth from the North to the South pole and extending it into space.

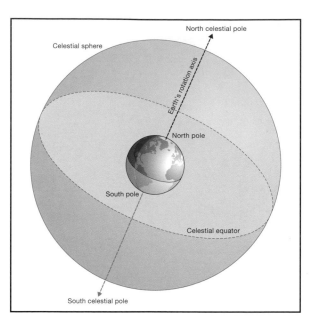

disappearance of some of the fixed stars matches the seasons. In the northern hemisphere, the appearance of the group of stars now known as the Orion constellation heralds the start of winter. Its disappearance is a sign that warmer weather and more plentiful food supplies are on the way. Just as the path of the Sun across the sky over the course of a day could be used to measure time, so the phases of the Moon could be used to track a longer period – a lunar month. The rising and setting

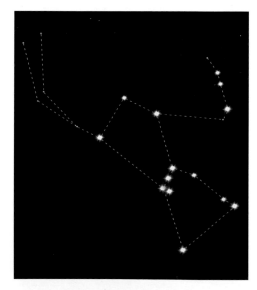

The constellation Orion is visible in the northern hemisphere in winter and in the southern hemisphere during summer.

positions of the Sun and some of the fixed stars could be used to track the course of the year. The first human uses of astronomical observation were almost certainly to keep track of time.

Our earliest ancestors tracked the movement of the Sun, Moon, planets and stars and learned how to predict and interpret them, using their knowledge to plant crops at appropriate times and to anticipate events such as annual floods or rains. But they probably also endowed the heavenly bodies they watched with supernatural significance.

Day to day

The oldest 'calendars' are vast archaeological sites that aligned posts or megaliths (giant stones) with the rising of the Sun or Moon on significant dates, such as the summer or winter solstice. The earliest site so far discovered is Warren Field near Crathes Castle in Aberdeenshire, Scotland, found in

BABIES BY ORION

A piece of carved mammoth ivory discovered in a collapsed cave complex in Geißenklösterle, Germany, is thought to be the earliest depiction of an asterism – a pattern of stars. Just 14cm (5½in) long, the carving is 32,000–35,000 years old. One side shows a human or part-human figure, taken to be Orion. The other side shows a series of pits and notches. It has been suggested that the pattern acts as a calendar which could be used to time the conception of a baby. If conception coincided with the arrival of Orion in the sky over Palaeolithic Germany, the baby would be born at a time when mother and infant would benefit from the food and warmth of summer for three months before winter set in again.

2004. It comprises 12 pits arranged in an arc. At least one pit held a post at some time. It seems likely that the monument performed some sort of calendric function. Archaeologists propose that the pits were used to track the lunar cycle, keeping a record of the lunar months. One of the pits (number 6)

also aligns with the position of sunrise at the winter solstice 10,000 years ago.

One suggestion is that the pits at Warren Field might have been used to tally lunar periods over the course of a

The phases of the Moon over a full lunar month.

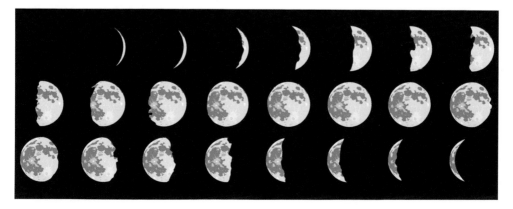

LUNAR AND SOLAR CALENDARS

Time is naturally divided astronomically by the Earth's orbit around the Sun (a year), the Earth's rotation (a day) and the phases of the Moon. A lunar month (a full cycle from one new or full moon to the next) is approximately 29½ days long. A year is 365¼ days. Inconveniently, a year is 12.37 lunar months long. For early societies, a lunar month was a useful and countable period of time, one that could be easily observed and checked just by looking up at the night sky. But if you use twelve lunar months as the basis of your year, the calendar will drift out of sync quite quickly. It will be a month out after only three years, and six months out after 18 years. To avoid this, an extra (intercalary) month has to be added every few years.

The earliest structures built to act as calendars, such as Warren Field, seem to be designed to help calibrate the solar year and lunar months. This can be done by picking a day – the winter solstice is the most convenient as it has the longest night – and noting the moon phase on that day. When the same phase next occurs at the solstice, it's time to add an intercalary month to keep lunar and solar calendars in sync. So if, say, we began a calendar with the winter solstice starting on the day of a full moon in Year 0, a full year – 12.37 lunar months later – the winter solstice would fall about a third (precisely 0.37) of the way through the 13th lunar cycle. The following year (Year 2), it would fall 0.74 of the way through the 13th lunar cycle. The next year (Year 3), it would be at the full moon again, but a total of 37 lunar cycles would have passed. If you were naming the months January to December, you would have got to the end of the fourth January by the time the third year had passed. To avoid starting the new year with February, you would need to add an extra month to the year just ending.

year. The priest-astronomers marked each lunar month as it passed, perhaps dropping a stone into a pit or moving a post to the next pit. Reaching the final pit meant the end of the year, and they would start again with the first pit. The midwinter solstice could be used to recalibrate. Each time the winter solstice fell at (say) a full moon, they added an extra month to the ending year. This would happen once over three years (see box, above). That the winter solstice was marked by the middle pit suggests that it came in the middle of the year for the people who used it, meaning their year started in late June (at the summer solstice).

The site seems to have been modified, apparently to adapt to shifting astronomical positions over a period of 6,000 years. The modifications suggest it was used continuously during that time.

As far as we know, Warren Field was a unique structure – it is 5,000 years older than any other known calendar-monument. But it could just be that others haven't yet been found. After all, Warren Field was only discovered in 2004 and its significance remained obscure until 2013.

The next oldest calendric structure is the Goseck circle, in Germany, constructed around 4,900 years ago – so only half the

SOLSTICES AND EQUINOXES

Standing on Earth and gazing out to space, it looks as though the Sun goes round the Earth against a background of stars. Astronomers call the path the Sun follows over the course of a year the 'ecliptic'. If we project the Earth's equator onto the sky, calling it the celestial equator, the Sun will seem to be above the celestial equator for half the year and below it for the other half of the year. There is a difference between the celestial equator and the ecliptic because the

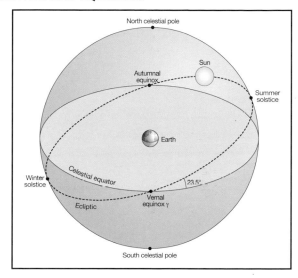

Earth tilts on its axis. The tilt is 23.5 degrees, and this tilt gives us the seasons, with days of different lengths.

For early societies, important days in the natural (solar) year were the summer and winter solstices and the spring (vernal) and autumn (autumnal) equinoxes. The solstices fall in

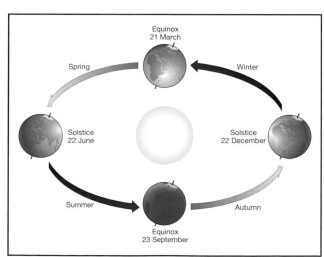

December and June, when the Sun is at its furthest point from the celestial equator. The longest day and night occur at the solstices. The equinoxes fall in March and September, when the ecliptic crosses the celestial equator. Day and night are of equal length at these two points (the word equinox means 'equal nights').

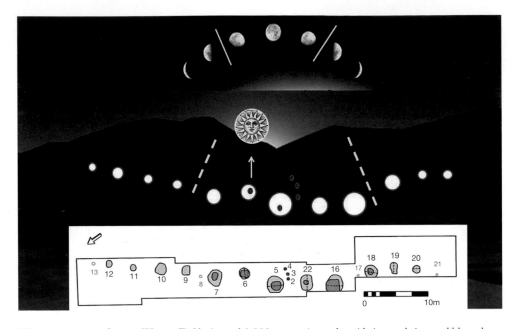

The arrangement of pits at Warren Field. Around 8,000BC, sunrise at the midwinter solstice would have been at the pass between the two central hills.

age of Warren Field. There are many more circular and elliptical structures in Central Europe, ranging through Poland, Germany, Austria, Slovakia, Hungary and the Czech Republic. All were constructed over a period of about 200 years, some 5,000 years ago. Like most of the other later

sites, Goseck allows the winter solstice to be determined from the alignment of sunrise and sunset. The site comprised four

The Goseck circle in Germany. The entrance points at bottom left and right show the direction of sunrise and sunset at the winter solstice, converging at the circle's midpoint.

STICKS AND STONE: STONEHENGE

Stonehenge is a large stone circle in Wiltshire, England, comprising a series of upright stones that originally supported horizontal stones (lintels). Some of the lintels remain in place. The uprights and lintels are made of bluestone and sandstone, the latter mined locally but the former hewn from hills in Wales and transported 250km (155 miles) to the site by land and/or water. The largest stone, the heel stone, weighs 30 tons. There is also an altar stone made of red sandstone. It is the most sophisticated stone circle in the world.

The first monument at Stonehenge was a circular earthwork enclosure – a ditch containing a ring of 56 timber or stone posts. This was built around 3000BC. It was used as a cemetery; cremations were carried out there for several centuries. The stone monument was built around 2500BC. Stonehenge is part of a complex of sites that were used continuously for around 2,000 years.

concentric circles, made up of a central mound, a surrounding ditch, and two wooden palisades. Gates in the palisades faced southeast, southwest and north. At the winter solstice, the rising sun aligned with the southeast gate and the setting sun aligned with the southwest gate.

Alignments rediscovered

Prehistoric astronomers left no user manuals for their calendric monuments; their uses had to be rediscovered by archaeologists with knowledge of astronomy (archaeoastronomers).

The notion that ancient monuments might be lined up with astronomical landmarks (or skymarks) first surfaced in 1909, when the eminent British astronomer Norman Lockyer (1836–1920) proposed that Stonehenge had been built as an ancient observatory. Lockyer, famous for discovering helium (see page 120) and founding the journal *Nature*, noticed while on holiday in Greece that some ancient

Norman Lockyer was the first astronomer to explore archaeoastronomy, finding astronomical significance in ancient archaeological sites.

PRECESSION

Axial precession, also called precession of the equinoxes, is the gradual movement of the Earth's axis, which results in the slow shift of the apparent positions of the stars.

The Earth's axis is at a tilt, a feature which produces the seasons as the Earth orbits the Sun. Over time, the axis moves in a circular path (see image, below). As the axis moves, the apparent position of the stars slowly changes, as we are looking at them from a slightly different angle. The positions of the celestial poles change, too. Polaris is currently the Pole Star in the northern hemisphere, and will be in its best possible Pole Star position around AD2100. Around AD3000, Gamma Cephei will take over the role. Polaris will take its turn again around AD27,800.

temples had apparently been rebuilt. Closer inspection revealed they had also been slightly realigned. Rebuilding ancient temples is a lot of work, especially for a pre-industrial culture, and would not have been undertaken lightly. Lockyer realized that the reason must have been to align them with the Sun, stars or planets, making a correction for the shifting appearance of the sky over centuries (see box, above). Turning his attention to Egypt, he found buildings there that were also aligned with celestial markers. At last, he looked at Stonehenge and found it aligned to face sunrise at the summer solstice. Although many of Lockyer's conjectures are now rejected (that Stonehenge was built by immigrants from the Far East, for example), the alignment of Stonehenge is uncontested and that it might have been used as an observatory of some kind remains a possibility.

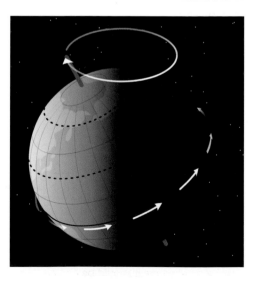

The Earth rotates on its axis (pole to pole) once a day; the axis itself rotates slowly, taking around 26,000 years to complete a revolution.

In the temple complex at Mnajdra, Malta (3600–3200BC), sunlight passes through the doorway at the equinoxes, lighting up the axial passageway and altar; at the solstices, it falls on the edge of megaliths standing to the right and left of the doorway and lighting up corresponding stones inside.

The Nebra sky disc

A much smaller circle, which contains the earliest known depiction of the cosmos, was found near the Goseck circle. It comes from the boundary between the prehistoric and historic eras in Europe. The Nebra sky disc is a bronze disc measuring 30cm (12in) across, dating from around 1600BC. It shows either the sun or the full moon, a crescent moon, and a collection of stars that represent the Pleiades as they would have looked 3,600 years ago. The crescent

---

CITIES AND STARS

Not just individual monuments line up with celestial phenomena. In 2016, a 15-year-old Canadian schoolboy, William Gadoury, combined his interests in Mayan culture and astronomy and made an astonishing discovery – the ruins of a vast Mayan city hidden in the jungle.

The Mayans occupied an area from southeast Mexico to Honduras and El Salvador, building cities from c.750BC. Gadoury noticed that if he projected 22 Mayan constellations on to a map, the stars in them corresponded exactly with the location of 117 Mayan settlements – a correspondence never noted before. He turned to a 23rd constellation and found that one of the three stars had no known corresponding city. Convinced there must be one, he worked out where it should be and contacted the Canadian Space Agency for help. With the use of satellite imagery through NASA and JAXA, the US and Japanese space agencies, the outlines of buildings buried in the rainforest emerged.

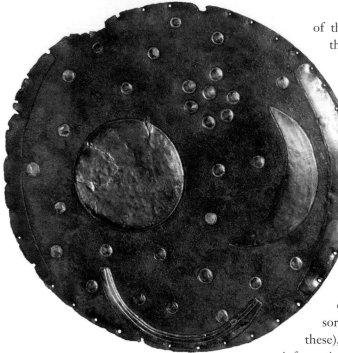

The Nebra sky disc is the earliest known portable astronomical tool.

is not a new moon, but a moon four or five days old. The astronomical features are in gold leaf, and the greenish background colour was achieved by applying rotten egg to bronze. It is a sophisticated artefact that someone has taken care and time to make. The disc was dug up by archaeological looters and passed around the German black market in antiquities for several years before it was seized in a police raid in 2002.

The use of the disc was initially obscure, but was explained in 2006 by a team of German researchers led by Harald Meller. He examined Babylonian astrological texts written about 1,000 years after the disc was made. These explain when to add an intercalary month by checking the position of the four-day-old moon against the Pleiades, just as the disc shows.

From prehistory to history

We can deduce what these ancient sites around the world were used for, but not the purpose to which the information gained was put. We can assume the function was for worship, divination, time-keeping, calendar-making for social or agricultural purposes, or sorcery (or a combination of all these), but the way in which the information was used is lost to us now. Perhaps it showed when to plant crops or move herds, or revealed propitious dates for personal rites of passage or when sacrifices or prayers should be offered.

With the beginning of written records – the start of the historical period – we start to see what our ancestors knew about the sky and how they used that information.

The first records

The Sumerians developed the first writing system in the world, around 3,200–3,500BC. Called cuneiform, it was made by pressing a wedge-shaped stylus into a clay tablet. Many clay tablets have survived, and some Babylonian astronomy can be pieced together by studying them.

The oldest astronomical text is tablet 63 of Enûma Anu Enlil, a large collection of omens related to astronomical investigations from the 2nd millennium BC when the Babylonians occupied Mesopotamia. Tablet 63 records the first and last risings of the planet Venus over a period of 21 years, and shows that the Babylonians were aware of

PLANET-RISE AND PLANET-SET

Because each planet is on its own orbital course around the Sun at the same time as the Earth is orbiting the Sun, its position varies in relation to us. Sometimes a planet is obscured by the Sun, and sometimes it is directly between Earth and the Sun and therefore lost in the Sun's glare. In between, it is seen to rise (appear above the horizon) either in the early morning (if it has just passed in front of the Sun) or set in the evening (if it has just passed behind the Sun).

When planet-rise is visible, the planet is not seen to set but just fades into the daytime sky, and when planet-set is visible the planet has risen during the daytime when it is invisible. The time at which the planet rises or sets moves further away from sunrise or sunset until it next disappears.

The planets and Moon follow the approximate path of the Sun along the ecliptic (see page 15). Just as sunrise and sunset happen at different points along the horizon, so do the rising and setting points for each planet. The interval from the start of a planetary cycle to its end (its return to the same point) is called the synodic period. For Mercury, that is 116 days and for Venus it is 584 days. This is not the same as the length of the planet's orbit, which is called its sidereal period.

ANOTHER CANDIDATE

The exact date of Enûma Anu Enlil is not known, but it's possible that the oldest astronomical record is not represented by Babylonian clay tablets at all but by an oracle 'bone' – a piece of carved turtle shell – from China. It records an eclipse of the Sun that took place on 5 June 1302BC, in the middle of the period during which Enûma Anu Enlil could have been created. It mentions 'three flames at the Sun, and big stars were seen'. Oracle bones were used for asking questions of the gods, so this was recorded in a context of superstition and religious belief rather than scientific interest.

the periodicity of planetary movements. The text was probably compiled in the Kassite period (1595–1157BC), but based on an earlier version or prototype. It includes names given to the stars by the Sumerians, who occupied the land before the Babylonians.

The most famous and significant Babylonian text is known as MUL.APIN after its opening logograms, meaning Plough Star. (By convention, Sumerian logograms are shown in capital letters, separated by a full stop.) Preserved on tablets made in 686BC, though probably compiled around 1000BC, it is an expansion on and improvement of the earlier 'Three Stars Each' catalogues (see page 154) which simply list the stars, and is clearly based on better observations. It's apparent from the tablets that Babylonian astronomers had a theory of planetary motion, though there is not enough information to piece it together in detail.

Priests and astronomers

What is clear from the presentation of information about Venus in the context of omens is that the Babylonians used astronomy for an astrological purpose. Those responsible for compiling and using the tablets were priests. There was an increase in astronomical activity during the reign of Nabonassar, or Nabû-nāṣir (747–734BC), with records of ominous events and detection of the 18-year cycle of lunar eclipses. Even though phenomena were being recorded and predicted for astrological purposes, the detail and accuracy with which they were observed led the Greek-Egyptian astronomer Ptolemy,

This Babylonian clay tablet from the 7th century BC records the rising and setting dates of the planet Venus during a period in the 2nd millennium BC.

800 years later, to consider this to be the period in which astronomy started, with the first usable data. Later Babylonian tablets (350–50BC) show astronomers occasionally using geometrical methods to calculate the position of Jupiter, but most of their

GEOMETRY OR ARITHMETIC?

The arithmetical method of calculating the positions of heavenly bodies works by collecting a lot of observational data, identifying patterns or averages and extrapolating into the future to make a prediction. For example, if a comet is observed at intervals of 50 years, an arithmetic prediction that it will appear in another 50 years can be made without needing to know anything about the nature of the comet's orbit.

The geometric method requires the astronomer to have a theory about the spatial relationships between celestial bodies. The prediction is derived from geometrical calculations based on those relationships. For example, we now know the orbital periods of the planets and their relation to Earth, so we can calculate their positions relative to Earth at future dates.

calculations, and all the earlier ones, were arithmetical and based on extrapolating from previous observations (see box, opposite).

Spreading it about

The astronomical knowledge amassed by the Babylonians formed the basis of Indian, Greek and possibly Chinese astronomy, the latter by way of India.

Although it's known that the Indus Valley civilization used astronomy to make calendars in the 3rd millennium BC, the people of the Indus Valley had no written language so we don't know the extent of their astronomical knowledge. The oldest surviving Indian astronomical text is the *Vedanga Jyotisha*. It explains how to track the movements of the Sun and Moon, which was important in organizing rituals. It survives in a copy from the 1st or 2nd century BC, but might have been composed around 700BC or later. Its origins lie much earlier: it describes the winter solstice of a date that probably lies between 1150BC and 1400BC. Vedanga Jyotisha resembles Babylonian work, suggesting that the Indian astronomers were familiar with Babylonian texts or methods, either at the time of its original composition or when the text was later revised.

Out on a limb

China has always been difficult to reach from the west, with the Himalayas and the Gobi Desert forming natural geological barriers. Further, isolationist policies cut China off from other centres of developing civilization. As a consequence, Chinese astronomy developed largely independently. No external influence can be traced before

the Three Kingdoms era (AD220–265) when translated Indian works on astronomy arrived in China along with Buddhism.

The Chinese developed their own accounts of the constellations and stars, with enduring star names that have been preserved on oracle bones from the middle of the Shang dynasty (c.1600–c.1046BC). The earliest known observatory in China, a carved platform 60m (197ft) across, was excavated in Shanxi province in 2005. It was used to locate the rising and setting of the Sun at different times of year, and dates from the Longshan period (2300–1900BC). Eclipses were recorded in China from around 750BC, providing an invaluable record for later astronomers. Detailed astronomical observations began in China during the Warring States period (usually given as 475–221BC), probably around 200BC.

Astronomy and astrology

The two principal practical uses of astronomy in the early historical period were time-keeping/calendar-making and location/navigation. But astronomy was also at the centre of religious and superstitious beliefs. Indeed, calendar-making and location often served these ends, too.

Today, scientists draw a very clear distinction between astronomy and astrology. Astronomy is a science, concerned with the movement, nature and history of bodies such as planets, stars, comets and asteroids. Astrology is superstition, a means of (supposedly) predicting events, interpreting the moods of the gods, and drawing correspondences between events in the heavens and events on Earth and in human lives.

It's easy to see how the idea of a link between earthly and celestial occurrences could come about. A culture observes that when a certain constellation rises above the horizon, the weather changes, and perhaps seeds grow. Whereas today we account for both the change in season and the appearance of the constellation by reference to the Earth's progress on its orbit around the Sun, a naïve observer could easily think the appearance of the constellation was responsible for the growth of seeds. It's a small step from there to worshipping the constellation, or beseeching it to intervene if seeds do not grow.

The most potent events in astrological terms were those that could not be as readily predicted as the rise of a constellation or the position of a planet. The appearance of a comet, a shooting star or an eclipse was taken to signal extraordinary events either already happening or about to happen. It could herald a great event that would be celebrated or a catastrophe to be feared.

The world and the stars

The earliest known system of astrology was Babylonian, developed around 1800BC. It might have been based on an earlier Sumerian system, but there is insufficient evidence to be sure. Babylonian astrology was of a type now known as 'mundane' astrology – it dealt with the fates of nations, cities, states and cultural leaders rather than with individuals. Modern astrology is more often associated with the personal – individual horoscopes and analyses of personality. The tablet-text Enûma Anu Enlil consists of 7,000 celestial omens recorded on 70 tablets.

Babylonian astrology linked gods with the planets and some of the fixed stars, and took the predicted or observed consequences of a planet's behaviour as an indication of the corresponding god's mood. It did not, though, link this directly with previous human action. Humans might try to sway or appease a god/planet in the case of an ill omen, but it was not assumed that the presaged event was intended as a punishment for humans. The nature of celestial omens often rested on previous events. So, if a new moon on a day of torrential rain was followed by a good event – an abundant harvest, perhaps – that pairing of new moon and rain would be considered auspicious next time it occurred.

Although rudimentary astrology started early in Babylon, it was only after Nabonassar became king in 747BC that it flourished and became more sophisticated. Prior to this, astronomical knowledge was sparse, so there was little to work with. Astronomers were not adept at predicting the movement of the planets, and an inability to predict normal behaviour makes it very difficult to detect the abnormal.

Patterns and a zodiac

Even the earliest astrologers depicted groups of stars that they considered went together. Cave paintings in Lascaux, France, made 17,300 years ago seem to show the Pleiades and Hyades star clusters. Seeing pictures in the combinations of stars, a sort of celestial join-the-dots activity, also predates written history.

Babylonian and Egyptian astronomers described pictures in the patterns made by the stars, and some of their asterisms have come down to us through Greek and Roman astronomy. Astronomers now distinguish between asterisms (pictures made from the patterns of the stars) and constellations (areas of the night sky), but in popular usage 'constellation' is used for both.

Some of the zodiacal asterisms in use now have survived since Babylonian times: Taurus, Leo, Scorpio, Sagittarius, Capricorn, Aquarius and possibly Virgo and Aries seem to be depicted on stones from Mesopotamia dating from the 14th century BC. The Babylonians were the first to divide a circle into 360 degrees, and to originate the zodiac by dividing the circle of the ecliptic into twelve 'houses', with each assigned 30 degrees of arc and an asterism, though the latter was probably perfected in the 6th century BC, after the end of the Babylonian empire. It was a small step from this arrangement of the sky to seeing some special relationship between the Sun and whichever constellation lay behind it, and this was something that astrology was quick to exploit.

The Babylonian zodiac was adopted, with modifications, by the Ancient Greeks through the work of Eudoxus of Cnidus in the 4th century BC. Its current version was devised by the Greek-Egyptian astronomer Claudius Ptolemy in the 2nd century AD. The earliest depiction of the circular zodiac dates from around 50BC and is the Dendera zodiac, an image found in (and sacked from) an Egyptian temple by Napoleon's expeditions into Egypt in the early 19th century.

The division of the sky into the 12 houses of the zodiac gave early astronomers a relatively easy way of seeing how the planets moved against the backdrop of the stars. The belief that the position and alignment of the planets influenced events on Earth was widespread, but the difficulty of working out exactly what the impact of planetary movements would be

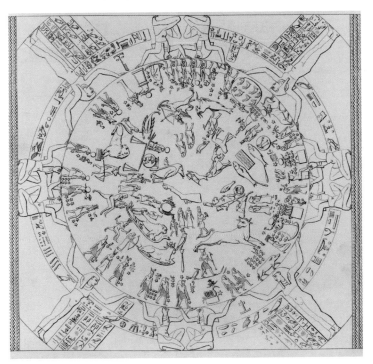

A sketch of the Dendera zodiac carved on the ceiling of the Temple of Hathor, Dendera.

The Mesoamerican site of Zempoala, near Veracruz in Mexico, has three mysterious stone circles composed of pillars. The circles have 40, 28 and 13 pillars. It's possible that they were used to keep track of eclipse cycles 1,000 years ago.

meant that it remained the province of expert astrologer-priests.

The heavens out of kilter

The reliable heavenly bodies bring a sense of security and predictability. Imagine, then, how terrifying or bewildering it would be when the sky was disrupted: when a new star appeared, when the sun was eclipsed by the moon, or when a star shone brightly and then vanished, never to return.

From time immemorial, people have feared unusual celestial events. In order to detect anything unusual or untoward in the

MOVING THROUGH THE ZODIAC

To understand how the houses of the signs of the zodiac work, imagine sitting on a fairground carousel and looking out as it moves to the scene beyond. You might see other stalls, the car park, the road – as the carousel turns, you see these in turn, one after the other, again and again. Now scale this up to the solar system, with the Earth as your place on the carousel and the stars forming the scene. As the Earth goes round the Sun, a different area of the starry background is visible. If we divide the full circle of the Earth's orbit into 12 sections we can pick out a star pattern for each section. The patterns that have been picked are the signs of the zodiac.

heavens – which might be a harbinger of doom – it was necessary to have a good grasp of the normal state of affairs. Only when astronomical observation reached a certain level of proficiency could astronomers and priests identify the truly remarkable. It took even more expertise and experience to be able to predict the unusual, and some events could never be predicted.

Observing the heavens for signs of trouble brewing on Earth was particularly important to the Chinese emperors. The collected data carefully gathered by sophisticated and thorough Chinese observations was put to largely astrological uses. Emperors reigned by virtue of the 'Mandate of Heaven'. The basic principle was that the Heavens granted a single emperor the right to rule, dependent on his personal virtue, and his mandate continued only as long as he ruled well. No family had the right to perpetual rule. Any signs that the emperor was not doing well and had lost the Mandate of Heaven meant he could be overthrown. The changes in dynasty that mark the periods of Chinese history are the points at which an imperial family has lost the Mandate of Heaven. Such a crisis was often signalled by disasters such as earthquakes, peasant uprisings

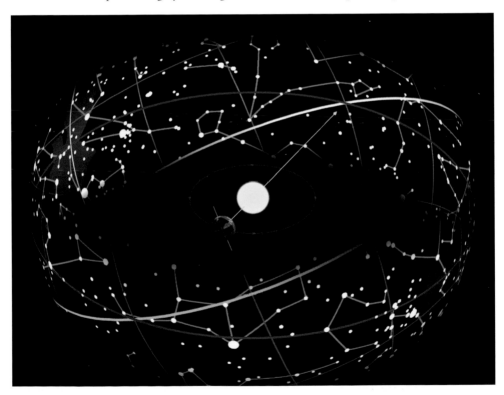

The constellations of the zodiac represent groups of stars that can be seen looking out from Earth along the path of the ecliptic – so looking away from the Sun towards deep space in the plane of Earth's orbit.

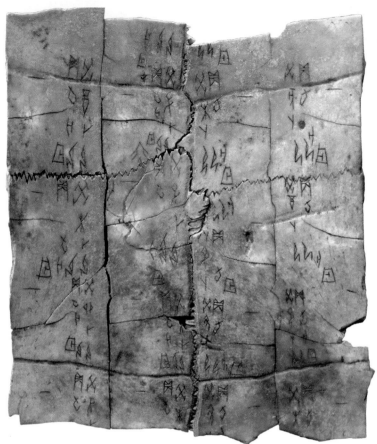

Oracle bones, in this case a fragment of turtle plastron, were used to question the gods. The question was inscribed on the bone, then the bone was heated and the pattern of cracks interpreted to give the answer. Astronomical events are sometimes recorded on oracle bones.

and famines. (Of course, these often went together, with a natural disaster leading to a famine, and hungry peasants staging an uprising.) Since disorder in the heavens mirrored or presaged disorder on Earth, keeping a careful watch on celestial events could, in theory, give an emperor warning all was not well.

Accordingly, emperors kept a retinue of astronomers, astrologers and meteorologists whose job it was to observe and interpret activity in the heavens. Five astronomers watched the sky from a platform, such as that in Shanxi province (see page 23), one looking towards each of the four points of the compass and one looking straight upwards. In the morning, they would report anything unusual to the Royal Astronomer who would interpret its significance and tell the emperor. A suitably unusual event might be the sighting of a comet (called a 'broom star' or 'long-tailed pheasant star' by the Chinese astronomers) or the occurrence of a lunar eclipse. Interpretation took the form of what corresponding events the emperor could expect to encounter on Earth. It was important work; numerous astronomers were executed for failing to predict or interpret events correctly. Unusual celestial events were consequently recorded with great care. This was necessary at the time because they were important prognostic tools, but it has been a boon to later astronomers, who now benefit from thorough records for more than 2,000 years and partial records going back 4,000 years.

While the Chinese were happy trying to see into the minds of the gods, astrology has not always been deemed acceptable. In the early days of Islam, astrology was condemned as being counter to the teaching that God has control over human lives and the universe, so it is blasphemous or sacrilegious to suggest that the planets

THE SILK ATLAS OF COMETS
A drawing dating from about 185BC, discovered in Hunan province, China, in 1973, shows comets of various styles, drawn carefully to distinguish between different configurations of tail and head. The manuscript depicts 29 comets observed over a period of 300 years. Alongside each is an account of the Earthly events (mostly catastrophes) associated with the comet, such as the death of a prince or a period of drought. The Silk Atlas is the earliest surviving catalogue of comets.

Comets have long been considered heralds of disaster or misfortune. Their unpredictability made it easy to see them as a sign from the gods.

have any effect. On the other hand, the movement of the planets could also be seen as controlled by God. But that, too, could lead to problems. Ibn Sina (c.980–1037) argued in *Essay on the Refutation of Astrology* that the planets do have an impact on Earthly events and, as such, manifest God's great power, but it was impossible for us to work out what the impact would be in advance (and blasphemous to think that we could). At the same time, other aspects of astrology were considered a respectable part of the natural sciences – when it would be a propitious time for blood-letting, for example. Astrology could provide simple guidelines for general behaviour and activity

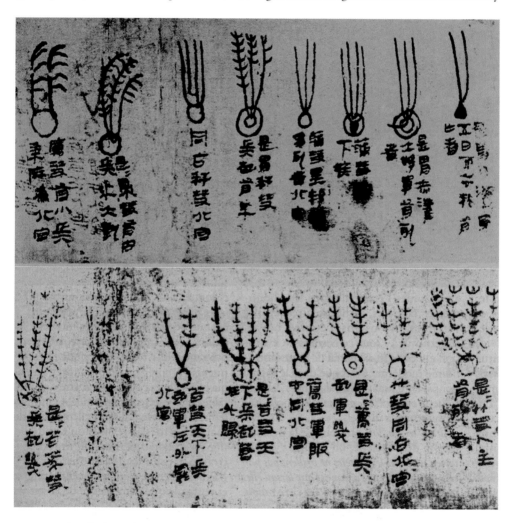

The Silk Atlas of Comets, from around 185BC, shows clearly the different types of heads and tails Chinese astronomers saw as they observed comets.

but could not suggest that the detail of lives could not be calculated in advance.

Counting the days

The Ancient Egyptians began their year with the heliacal rising of Sirius (which they called Sopdet). This is the first, brief appearance of the star above the horizon before sunrise. The first mention of this system dates to around 3000BC. The Egyptians divided the year into 365 days and three seasons: Inundation (when the Nile flooded, renewing the fertility of the land), Growth and Harvest. Because they did not take account of the extra quarter day per year, the Egyptian civic calendar slipped by a day every four years. The Sothic year (the time it took for Sirius to rise in exactly the same place) is almost exactly 365.25 days, the same as a solar year. Consequently, there were two calendars operating – one based on counting days and one based on observing Sirius – which looked pretty much the same in the short term, but slowly slipped out of sync over a longer period. The calendric year returned to the same heliacal rising of Sirius only once every 1,460 years. In 238BC, the Ptolemaic rulers of Egypt declared that every fourth year should have an extra day (which we consider a leap day), so bringing the civic calendar back into sync with the celestial calendar.

Other cultures developed independent calendars, notably China and the Mesoamerican civilizations. The origins of the Chinese calendar can be traced back to the 14th century BC. But legend has it that it was invented by Emperor Huangdi and started in 2637BC. By that time, the Chinese already knew that the year has 365¼ days and that a lunar month is 29½th days. Years were counted from the accession of an emperor (which always began a new era). A new era was seen as a chance to re-establish the connection between heaven and Earth. An emperor could also declare a new era, rebooting the connection, as it were, if

The star Sirius is represented by this Egyptian statue of a goddess.

A STABLE STAR

Sirius is unusual among the fixed stars in that it does not precess (see page 18). This made Sirius an especially suitable star for the Ancient Egyptians to use as the basis of their calendar. For them to have known this, Egyptian astronomers must have observed Sirius over some considerable time.

things had gone awry – signalled by a natural disaster or by astronomers failing to predict a celestial event such as an eclipse. The *yin-yang li* traditional calendar (literally 'heaven-Earth' calendar) was used, sometimes alongside imported calendars such as the Hindu calendar, until 1912 when China officially adopted the Western Gregorian calendar.

The development of accurate calendars was often driven by the need to fix religious festivals and observance, an impulse that continued with the formation of new religions. Both Christianity and Islam have put astronomy to use in this way. Arab astronomers and engineers were zealous in their pursuit of improved methods of keeping time so that daily prayers could be recited by the devout at the right time. Time-keeping on a larger scale was essential in scheduling religious festivals.

Place and navigation

Just as the great earthworks and stone circles of neolithic Europe were aligned with sunrise at the equinoxes, so later cultures aligned more ornate constructions with celestial markers.

The Mayans had no sophisticated astronomical tools or instruments, but through detailed, careful observation learned a great deal about the movements of celestial bodies and were able to make astonishingly accurate predictions. Many of their buildings aligned precisely with the equinoxes and midsummer, or the most northerly and southerly risings of Venus. It's likely that the astronomical alignment of buildings, and even entire cities, served a religious or superstitious purpose in many cultures.

Fixing locations by astronomical means was also used by literate cultures and drove the development of astronomy in some cases. Arab astronomy, in the service of Islam, sought to find ever more accurate ways of determining the direction of Mecca. The mathematician al-Khwarizmi (c.780–c.850) constructed a table of the latitudes and longitudes of 2,402 cities and landmarks that could be used by the faithful to help them to pray facing Mecca.

EASTER AND ASTRONOMY

The Catholic Church fixes the date of Easter, its celebration to mark the resurrection of Christ, using a method set out in AD325 by the Council of Nicaea. In the first centuries AD, Easter was celebrated on different days by different groups of Christians; the Council of Nicaea sought to standardize it.

Easter is now celebrated on the first Sunday after the first full moon occurring on or after the spring equinox. Obviously, early Christians couldn't simply wait to find out when that full moon would fall, then quickly celebrate Easter. For one thing, they had to fit in Lent – 40 days of fasting – immediately beforehand, so had to know several weeks in advance when that full moon would fall, a task that could only be achieved by keeping astronomical records and projecting into the future.

It's likely that the position and alignment of many structures, and even entire cities, had astronomical religious or superstitious significance in many cultures.

Finding your way

Using the stars to find a location is particularly important for voyagers, and most especially those travelling by sea when no landmarks are visible. The Māori made their way from eastern Polynesia to New Zealand, perhaps around 1280–1300, using only the night sky, weather patterns, wave patterns and ocean currents to navigate. Polynesian navigators used wave patterns and sea currents as well, but also relied (and still rely) on a 'star compass' – not a physical object but a mental construct that divides the horizon into 32 directions which correspond with the rising and setting positions of bright stars and the Pleiades group. Modern aeroplane pilots are still trained to navigate by the stars in case of emergency.

The Polynesians sailed north and south of the equator, so saw elements of both northern and southern skies. They learned which stars were directly overhead on the islands they occupied and visited, and could then locate the latitude of those islands by the positions of the stars. Once in the right area, they could look out for birds, driftwood, floating weeds, fish and local sea currents to guide them to landfall. Before them, Phoenician navigators plied the seas 4,000 years ago, using the Sun by day and the stars at night to plot their course.

It is quite easy to determine latitude – your position north or south of the equator. In the northern hemisphere, fixing the position of the North Star or measuring the angle of the Sun at noon above the horizon in degrees gives your latitude. This method

33

has been used for centuries. Longitude, your position east or west of your home port, is much trickier to determine. A reliable method was not discovered until 1765. For this reason, most seafaring people hugged the shoreline or hopped between islands rather than striking out boldly across vast expanses of ocean.

And so to science

In Mesopotamia, Egypt, India, China and possibly Mesoamerica, astronomy served a practical purpose of calendar-making and was difficult or impossible to separate from astrology. However, the Ancient Greeks took a different approach. They were the first people known to have proposed that the workings of the natural world – indeed, the whole universe – might be susceptible to investigation and explanation that did not involve a supernatural being. They questioned whether there might be no controlling hand; whether things might just be as they are because that's how matter and maths work. They were also the first actively to pursue knowledge for its own sake, not simply for its potential applications. Greece was the birthplace of science.

The first scientist

The philosopher Thales of Miletus (c.624–c.546BC) was the first to make pronouncements about the nature of the universe that did not rely on supernatural causes. As such, he is the forefather of the scientific method, laying the foundations

The Ancient Greek philosopher Thales initiated scientific thinking, seeking rational explanations for natural events and phenomena.

that made the great Western endeavour of science possible. It's unlikely that Thales took his view from any earlier or foreign tradition; he was lauded as an original thinker by those who came soon after him.

Thales had views on many subjects, including astronomy. In his cosmological model, the Earth floated in water, and it is likely that he considered the Earth to be spherical (though the Ancient Greeks did not all agree on this). His floating-Earth theory gave him an explanation for earthquakes – the Earth is tossed about on stormy seas. The prevailing, Homeric account was that earthquakes were caused by the god Poseidon striding angrily around and shaking the ground. It doesn't matter that Thales was wrong about the cause of earthquakes; what matters is that he believed there to be a purely rational explanation with no need to invoke supernatural beings or causes.

ΘΑΛΕΣ

Thales is also credited with discovering the precise dates of the solstices. They are hard to pinpoint, as the sun seems to stand still for a few days around the summer and winter solstices. More than 700 years later, Ptolemy acknowledged the problematic nature of the task. Thales would have had to observe sunrise and sunset over many days at the end of June and December and over several years to be able to predict the solstice accurately.

The Tal Qadi Stone, found near Salina in Malta, dates from the 4th millennium BC. It seems to show a pattern of stars and a crescent moon. If it was once a full circle, the stone divides the sky into 16 segments.

Greeks outside Greece

Greek (Hellenistic) culture wasn't restricted to the Greek mainland and islands as we now know them; 2,600 years ago, there were Greek colonies in Turkey, all around the coast of the Black Sea, the northern coast of Egypt and the south of Italy. Thales was a Milesian, from the city of Miletus in Turkey. The Greeks picked up the astronomy and mathematics of the Babylonians and Egyptians and made them their own. But unlike their predecessors, the Greeks took the facts and rejected the superstition.

Perhaps the moment that marks the transition from superstition to science in the history of astronomy came on 28 May 585BC. The Medes and Lydians were battling for control of Turkey. Thales had learned sufficient Babylonian astronomy to have been able to predict the total eclipse of the Sun that took place on that day. Consequently, the Milesian troops were prepared and unafraid when the eclipse occurred, but the Lydians were terrified and eager to make peace. Science, one point; superstition, nil. With the Greeks, a whole new chapter begins.

'On one occasion [the Medes and the Lydians] had an unexpected battle in the dark, an event which occurred after five years of indecisive warfare: the two armies had already engaged and the fight was in progress, when day was suddenly turned into night. This change from daylight to darkness had been foretold to the Ionians by Thales of Miletus, who fixed the date for it within the limits of the year in which it did, in fact, take place.'

Herodotus, *Histories*, c.425BC

CHAPTER 2

The great
SCHEME OF
THINGS

'He who does not know what the world is does not know where he is, and he who does not know for what purpose the world exists, does not know who he is, nor what the world is.'

Marcus Aurelius,
Roman emperor,
ruled AD161–180

Is the universe infinite or bounded, unchanging or changing? Is it a single universe through time or a cycle of creation and destruction? Should we think in terms of a single universe or are there multiple universes? We see the same basic models recurring throughout history, whether articulated through a collection of mythic stories or scientific theory.

Angels turn the cranks that rotate the outermost sphere of the Ptolemaic model of the universe, imparting motion to the celestial bodies (14th century).

Life, the universe, and everything . . .

Cosmology is the study of the entire universe, its structure and processes; there can be no bigger subject. It involves constructing models of how the universe might work and, latterly, testing those models against what we can observe and deduce. The earliest models could be tested only against how the stars, planets, Sun and Moon seemed to move relative to the Earth and what could be experienced on the Earth (the seasons, for example).

More recently, cosmological models have been tested through mathematics, through astronomical observation using advanced technology, through data and samples collected by space probes, and through their potential for integration into a complete, coherent and consistent system. As more is discovered, the model is challenged and adapted. Through a process of refinement, and occasionally of radical overhaul, the prevailing cosmological model has come closer to offering a full explanation of how the universe works.

Magical and mythical

Looking up at the sky and out across the land, the idea that the Earth is flat and the sky forms a dome over it is intuitive. That the clouds are lower than the sun, moon and stars is obvious, because they move quickly and can obscure the heavenly bodies. But nothing of the origins of the universe or the nature of the celestial bodies can be garnered simply by looking at the sky with the unaided eye. In ancient times, these were mysteries that were ripe for imaginative speculation.

The ceiling of the sarcophagus hall of the tomb of the Egyptian pharaoh Seti I, c.1279BC, showing the stars and constellations recognized in the 19th Dynasty with their representative deities.

From the earliest written records, a variety of cosmological models emerge, many supporting an extensive and complex mythology. These early accounts of the universe made no attempt to work out what it was actually like, but were inextricably entwined with mythological and religious accounts of origins and with narratives about the exploits and rivalries of gods. Certain motifs recur, such as the idea of a realm of gods above the Earth and a realm of the dead or evil spirits below the Earth. These are of more interest to anthropologists and historians of religion than to astronomers.

Layers of heaven and Earth

As far as it is possible to piece together the cosmology of the Sumerians (and later the Babylonians), it seems to propose a relatively flat, circular Earth under dome-shaped heavens. Together they formed a finite, sealed unit floating in the infinite waters of chaos.

The ancient Sumerians considered the firmament a solid shell which rested on the edges of the Earth. The Earth had some measure of thickness, as it's possible to dig down into it and to find underground caves. The underworld was also contained within the Earth. The stars, Moon and Sun were within the sealed Earth-heavens composite.

The Babylonians took on the same basic cosmological structure but made it more complex and embellished the mythology behind it. The Babylonian universe had at least two levels of heaven above Earth and one cosmic region, Apsu, below, but in other versions there were six levels, three for heaven, one for Earth, one for the underworld of the dead and one cosmic region separating the realms of living and dead. In both Sumerian and Babylonian accounts, the regions are closely associated with gods and mythological narratives rather than prompting any attempt to account for observations. The sun travels along a river in the day from east to west, then spends the night in an underground tunnel, returning to the position from which it will rise again.

The Babylonian cosmological model, with the heavens arching over the arched (brown) Earth.

The overarching sky goddess

Ancient Egyptian cosmology shared a lot with contemporary ideas in Mesopotamia. The Earth was considered rectangular rather than round, but still flat. The Nile ran down the centre of it. The sky was again held up by columns, but these columns were now the arms and legs of the goddess *Nut*. According to one of many versions of the Egyptian origin myth, Nut was embracing her husband *Sibû*, the earth god, when another god, *Shû*, grabbed Nut and lifted her up to become the sky. Shû froze Sibû as he was thrashing about in protest, and his twisted form accounts for the uneven surface of the Earth.

Nut gives birth to the Sun god, Ra, each day. He passes over her body, which arches over the Earth, and is swallowed by her each night, ready to be reborn the next morning.

This is only one of many versions of the mythical relationship between Nut and Ra.

Chinese cosmology

There are three distinct traditions in Chinese cosmology. The earliest recorded was described in the 3rd century BC in *Master Li's Spring and Summer Annals*. It describes the heavens as a hemispherical dome, encasing a domed Earth. The distance between the two was thought to be 80,000 li (around 43,000km/26,700 miles). The north celestial pole was directly above the centre of the Earth (China, of course), and the heavens revolved around them.

The next was the celestial sphere model, associated with Loxia Hong (d.104BC) and described later by the great Chinese astronomer Zhang Heng (AD78–139): 'The

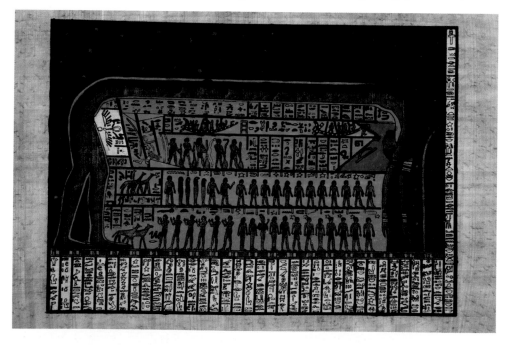

The Egyptian goddess Nut arches over the Earth, forming the sky.

TURTLES ALL THE WAY DOWN

A cosmological model associated with the author Terry Pratchett's *Discworld* series is that of a world supported by four elephants who in turn stand on a giant turtle. It seems to have its origins in Hindu mythology, though the earliest reference to a Hindu source is in 1599. The 'world turtle', known as Akupara, supports seven (or four) elephants that hold up the half-sphere of the world.

sky is like a hen's egg, and is as round as a crossbow pellet, the Earth is like the yolk of the egg, lying alone at the centre. The sky is large and the Earth small.'

The Chinese Xuan Ye tradition, on the other hand, considered the heavens to be infinite, with celestial bodies dotted around at intervals and each of them able to move, driven by celestial winds. Associated with Qi Meng, who lived some time in the first two centuries AD, it is described in a text written in the 4th century by Ge Hong: 'The sun, moon and company of stars float freely in empty space, moving or standing still, and all of them are nothing but condensed vapour. The seven luminaries [Sun, Moon and five known planets] sometimes appear and sometimes disappear, sometimes move forward and sometimes retrograde, seeming each to follow a different series of regularities. Their advances and recessions are not the same . . . they are not in any way

tied together. The Pole star alone keeps its place. . . . The speed of the luminaries depends on their individual natures.'

After AD520, the domed hemisphere model was predominant.

Space for thought

The notion of a cosmological model as a starting point for investigation instead of an end point is a relatively modern one.

There is a great difference between *making up* a story of gods and goddesses (or believing one passed down) and trying to *work out* what might be true by extending what we can observe on Earth. The Sun is hot and bright, as is a fire – so perhaps the Sun is a fire in the sky? This type of conjecture is based on reason, not on belief. Why does the Sun not fall down? Perhaps because it is held up by an invisible god – or perhaps because it is somehow fixed to the dome of the sky so that it can't fall. The former is a mythical explanation, which is in one sense a way of avoiding explanation. The latter borrows from the familiar – we fix a lamp to the wall so that it can't fall and set fire to the house, so perhaps this would work in the heavens, too. If the Sun is fixed to the dome of heaven, how is it fixed and what is the dome made of? Questions proliferate. But if it is held up by a god, well, we know nothing of gods and why they might bother to do that, so we can just leave them to it. Science doesn't accept that something is unknowable, and therefore unworthy of investigation; only that it is not known yet, but might be explained in the future. It gives us room to think.

Starting to be scientific

Separate mythological and proto-scientific cosmological traditions began to develop in Ancient Greece. In Greek mythic cosmology, Gaia was the goddess of the Earth and mother of all creation, producing numerous progeny from her various unions with the sky, the sea and the stormy pit of Tartarus beneath the Earth. This mythical cosmological model, like that of the Babylonians, had a flat Earth covered by the dome of the heavens. There was a corresponding dome under the Earth, which formed the pit of Tartarus where the Titans were imprisoned after their defeat by the gods.

While this fabulous narrative continued to have currency, scientifically minded philosophers began to formulate an account which had no roots in the supernatural. Around 500BC, the philosopher Heraclitus used the word *kosmos*, the root of our word cosmology, and rejected the idea of a divine origin of the universe: 'This world-order [kosmos], the same of all, no god nor man did create, but it ever was and is and will be: everliving fire, kindling in measures and being quenched in measures.'

Modelling the universe

Heraclitus was not the first to look for a rational explanation. Anaximander (c.610– 546BC) was the first speculative astronomer, metaphysician and geographer, making the earliest-known world map. Little of his work survives, but reports of his thinking were left by others, including the great philosopher and proto-scientist Aristotle (384–322BC).

Although Anaximander is not known to have made any astronomical observations, he made a crucial leap which kick-started Western astronomy. He put his efforts into speculative thinking, such as explaining the working of the cosmos, and made three key points:
- Celestial bodies move in full circles, so pass beneath the Earth as well as above it.
This is not entirely clear from observation, but seems likely since the stars near the Pole Star can be seen to trace full circles. It was a daring idea because it depends on a second point, which is that:
- The Earth floats unsupported in space.
This could not be demonstrated by observation until 2,500 years later, when space travel finally enabled us to photograph Earth from space. Anaximander believed the Earth to be a round column or thick disc, with a diameter three times its height. We live on the top surface. The Earth does not fall through space, he maintained, because it is in the centre; pressures in all directions are equal and the Earth has no reason to move

The Pre-Socratic Greek philosopher Anaximander was the first person to take a rational approach to astronomy.

Anaximander's explanation of the stars, planets, Moon and Sun was rather more fanciful. He suggested that the celestial bodies are like chariot wheels, each with a rim made of opaque vapour, but filled with fire. There are gaps in the rim through which the light can shine. While the Moon and Sun have a wheel each, the stars presumably have several wheels, each with more than one gap (a wheel for each star would make it rather crowded). The wheels are at fixed distances from Earth, giving a model of the universe as a central Earth surrounded by concentric circles made up of these celestial wheels. The wheels do not all move at the same speed, and the axis of the heavens is at a 38.5 degree tilt when measured at Delphi, the 'navel of the world'.

in one direction rather than another. (This is the first known use of the 'sufficient cause' argument – the idea that for something to happen there must be a reason.)

• The celestial bodies are not all on the same spherical plane, but lie behind one another.

This was entirely novel. Previously, the heavens had been described as a single shell or dome to the inside of which all the celestial bodies were attached. For the first time, Anaximander conceived of objects in space receding into the distance. Oddly, he put the fixed stars closest to Earth, followed by the Moon, with the Sun furthest away. These three concepts have been fundamental to all models of the universe since and can be said to form the foundations of modern astronomy. None could be tested empirically in Anaximander's time.

Anaximander's model of the universe, with the distances between bodies marked in multiples of the Earth's diameter.

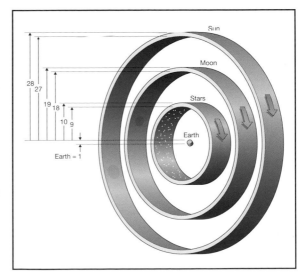

43

A BIT ABOUT MATTER

The universe is made of stuff, or matter as we now term it. The nature of matter was disputed in Ancient Greece. Empedocles (c.492–432BC) proposed that all things are composed of four 'roots' or elements: fire, air, water and earth. The proportions in which these elements are mixed account for the characteristics of different types of matter. Aristotle added a fifth element, *aether*. It differed from the terrestrial elements in having no properties of heat or cold, wetness or dryness, and being present only in the circles of the celestial bodies.

A trickier question for the Ancient Greeks was whether matter is continuous or discontinuous – that is, whether or not it is divided into tiny portions. The notion that everything in the universe is composed of tiny, indivisible particles called atoms (or *atomos*, meaning uncuttables) was proposed by either the philosopher Leucippus (d.370BC) or one of his followers, Democritus (460–370BC). The qualities of different types of matter were thought to be the result of the different arrangement of the atoms. It was a pretty good guess, given that their work was entirely speculative.

Empty space?

The notion that most of space is empty is entirely familiar to us now. But the concept of completely empty space, a true void, was contentious for a long time. Leucippus is credited with first proposing a void in the 4th or 5th century BC. A void can exist only if matter is in tiny lumps (if matter is continuous, there is no space between it)

so the idea of a void necessarily goes along with atomism.

The void presented a philosophical problem. Plenty of philosophers and thinkers over the centuries have rejected the idea that there can be space containing nothing. Aristotle was one of them, arguing instead that matter is not comprised of separate atoms, but is everywhere and continuous. Where there is not something else, there is *aether*. The Ancient Greeks had no way of answering the question of the void. It lay unresolved for centuries (see Chapter 7).

Our place in space

Whether or not space is empty, the question of how celestial bodies are arranged in it became contentious and soon crucial. Standing on Earth, it is impossible to tell whether the Sun revolves around the Earth or the Earth revolves around the Sun – it would look the same, whichever way round it was.

Sitting round the hearth (of the universe)

Although Anaximander set the first scientific model around a central Earth, this was not the only model in Ancient Greece. The

> 'Just as a passenger in a boat moving downstream sees the stationary (trees on the river banks) as traversing upstream, so does an observer on earth see the fixed stars as moving towards the west at exactly the same speed (at which the earth moves from west to east.)'
>
> Aryabhata,
> *Aryabhatiya*, 6th century AD

followers of the mathematician Pythagoras (*c*.570–*c*.495BC) believed that the universe orbited a central invisible fire, sometimes called the Hearth of the Universe. The model was possibly first drawn up by Philolaus (*c*.470–385BC). Ten concentric crystal spheres centred on an invisible fire held, in turn, a Counter-Earth, Earth, the Moon, the Sun, the five planets in order and finally the fixed stars. Where these crystal spheres touch one another, they make a beautiful sound, the 'music of the spheres'. According to Aristotle, the Counter-Earth not only helped to explain eclipses but also made up the numbers – the Pythagoreans held 10 to be a sacred or perfect number and so would greatly have preferred a system with ten circles.

(Aristotle was no fan of this model and might well have been having a snide dig at the Pythagoreans.) The Counter-Earth might have been added to maintain balance and harmony, since otherwise there would be a single massive body orbiting the centre and the system would be perpetually lopsided. Although the Earth orbits the central fire, it does not revolve on its own axis in this model. This explained why neither the central fire nor the Counter-Earth was visible from Greece – Greece simply faced away from them.

The Pythagorean model might have been the first which did not put the Earth at the centre of the universe, but it was only displaced by two bodies that don't actually exist.

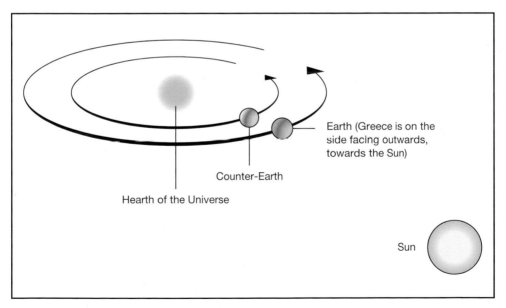

In the model of the universe attributed to Philolaus, the Earth, Sun and a 'Counter-Earth' all revolve around the 'Hearth of the Universe'. We can't see the Counter-Earth or the Hearth of the Universe because the Earth does not spin on its axis and we live on the side facing the Sun.

The centre of all things

Aristarchus of Samos (*c*.310–*c*.230BC) is the first person known to have argued that the Sun, not the Earth, is at the centre of the solar system, and so of the known universe. According to his model, the Earth rotates daily on its own axis and traces a year-long orbit around the Sun. The other planets and the fixed stars also occupy concentric circles or spheres around the Sun. Aristarchus further suggested that the stars are suns, but are too far away for us to feel their heat. He said that their great distance from Earth is the reason they seem to stay the same distance from one another. These are quite stunning insights for someone whose only resource was naked-eye observation of the night sky. The movement of the stars relative to one another can now be measured with telescopes and used to calculate the distance to the stars, but was not detectable at the time of Aristarchus.

Aristarchus's model didn't really catch on. There was no compelling evidence for one model over another, and human pride might have made the geocentric (Earth-centred) model preferable – humans like to be the centre of celestial attention. It also seems more intuitive: we see the Sun apparently move across the sky, and the fixed stars apparently revolve around the Pole Star. Why assume that they are not moving around the Earth?

The heliocentric (Sun-centred) view enjoyed intermittent support, though. Around 100 years after Aristarchus, Seleucus of Seleucia promoted it. His writings do not survive, but his work is known from records left by other writers. The Greek historian Plutarch (AD46–120) claimed that Seleucus

---

A SIMPLE MISTAKE

It's commonly said that Aristarchus's heliocentric model was ignored, and even that he was perhaps threatened for suggesting it, but the evidence suggests otherwise. The Roman natural scientist Pliny the Elder (AD23–79) and the playwright and philosopher Seneca (4BC–AD65) both refer to the *apparent* retrograde movement of the planets (see page 48). This suggests that they accepted a heliocentric model in which, although the planets appear to move backwards, this is a consequence of the Earth's movement relative to them and not the actual movement of the planets. The suggestion that Aristarchus was criticized or persecuted for his theory comes from a 17th-century mistranslation of Plutarch's account of Aristarchus.

---

was the first person to demonstrate by reasoning that the Earth moves around the Sun, though his argument is not preserved. Seleucus is also credited with being the first to deduce that the tides are caused by the influence of the Moon, and the first person to suggest that the universe is infinite.

Aristotle backs the wrong horse

Aristotle is acknowledged as one of the greatest thinkers of all time, but astronomy was not his strong point. Besides denying the possibility of a void, he rejected Aristarchus's account in favour of the generally accepted geocentric model. Since Aristotle was extremely influential, this set the scene for centuries, with the whole of the Western world setting off on the wrong track.

The geocentric model of the universe puts the Earth at the centre and has the Moon, Sun, planets and stars revolving around it.

In order to understand Aristotle's cosmology, it's necessary to know a little of his theory of physics. He accepted that earthly matter is made up of the four elements – earth, water, air and fire – but added a fifth element, *aether*, for the heavens. He divided motion into two types: natural motion, or compelled motion. The first described the natural motion of an element, which was in straight lines for the four earthly elements and in circles for the *aether*. Each of the terrestrial elements naturally moved towards or away from the centre of the universe (the centre of the Earth). Earth, being heaviest, moved downwards through the others; air and fire move naturally upwards. Compelled movement, such as when we throw a stone upwards, cannot be maintained indefinitely – eventually the stone will fall down, following its natural movement. For movement to continue, the moving object needs constant contact with the originator of the movement.

This contact can be delegated – so if you throw a stone, the force to move the stone after it has left your hand passes to the air, which moves out of the way of the stone, goes round the back, and gives it a push! It's not a very compelling argument, but it was acceptable to Aristotle. This is why, though, he could have no empty space. With nothing to keep exerting pressure on something moving, it would not continue to move – yet we can see that the heavenly bodies do continue to move.

When extended to cosmology, Aristotle's ideas led him to an Earth-centred universe with two distinct zones. The sub-lunar zone is made up of the earthly elements with their predominantly rectilinear movements; the heavenly bodies

Aristotle was one of the greatest thinkers ever to live, but in some fields – astronomy included – his mistakes held back progress.

are made of *aether* and have only circular movement. The Moon, Sun, planets and fixed stars then inhabited spheres travelling around the Earth in perfectly circular motion. This outer region was perfect, eternal and never subject to change. Consequently, phenomena such as meteors and comets were located by Aristotle in the upper atmosphere.

Problems with planets

This is a nice idea, but sadly it does not quite match observations. The planets do not move in steady circles around the Earth. Watching them over a period of a year, they seem to describe a series of loops; a planet will stop, then appear to loop backwards, then stop again before resuming its forwards movement. These periods of retrograde indicate that they do not follow a simple orbit around the Earth.

The Greek mathematician and astronomer Apollonius of Perga (262–190BC) came up with the start of a solution. He suggested that each planet's path is defined by two circles. The planet is in a small circular orbit, called an 'epicycle'. This epicycle is in orbit around the Earth. The larger circle traced around the Earth is called the 'deferent'.

Even this did not quite explain the movement, though, as the retrograde loops are not evenly spaced or of equal angular size. The next step was to offset the deferent so that the Earth was not at the centre of

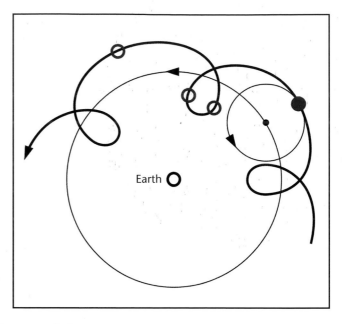

The planet (red) goes round in its own epicycle, which in turn orbits Earth.

the circle. This added a new philosophical problem: if the planets were not orbiting the Earth, their movement was not uniformly circular. It all seemed a bit stuck.

The Ptolemaic model

In the 2nd century AD, the Greek-Egyptian astronomer Ptolemy added an extra point that allowed at least the illusion of uniform movement, and devised a mathematical explanation for the movements of the celestial bodies around the Earth. His relatively small change ensured the supremacy of the geocentric model for centuries to come.

With the deferent offset from the Earth, it has as its central focus a point in space called the 'eccentric'. Ptolemy added another point, opposite the Earth and equidistant from the eccentric, which he called the 'equant'. The planet's speed was uniform in relation to the equant. This means that if you could stand at the equant and watch, the centre of the planet's epicycle would always move at the same angular speed (it would cover the same angle of arc in the same period of time). From anywhere else, including Earth and the eccentric, the planet is seen to go more quickly for some parts of its orbit than others. This restored the uniform circular motion that Aristotle demanded at the same time as explaining the apparent movements of the planets as seen from Earth.

Ptolemy put each of the heavenly bodies on its own sphere or orb, with the Moon closest to Earth, followed by Mercury, Venus, the Sun, Mars, Jupiter and Saturn. Finally, the fixed stars share the outermost sphere. There was no certain way to determine the sequence, except that the Moon must be closest as it can occlude all of the others, and the fixed stars must be furthest away as the others move in front of them. The sequence Mars, Jupiter, Saturn follows from their progressively long sidereal periods – the time it takes for them to move through a full zodiacal sequence. For Mercury, Venus and the Sun, which all

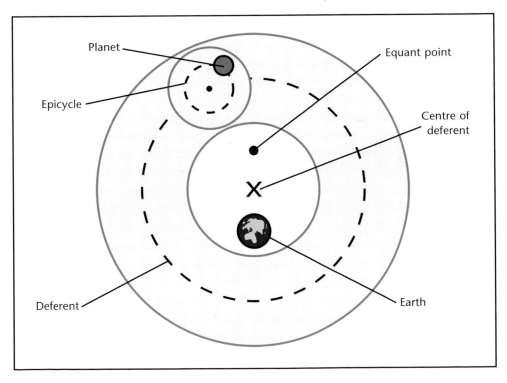

Ptolemy's explanation of the movement of the planets had each planet orbiting in an epicycle, the centre of which orbits the point X, the centre of the deferent. Earth and the equant point are equally offset from the centre.

CLAUDIUS PTOLEMY (AD c.85–c.165)

Relatively little is known about Ptolemy's life except that he lived in Alexandria, Egypt, and made astronomical observations in the period AD127–41. His most famous work is the 13-volume treatise known as the *Almagest* (from its name in Arabic, *al-majisti*). Ptolemy's great achievement was his explanation of the movement of the planets, giving mathematical methods and calculations that are apparently original to him.

The book first introduces the model derived from Aristotle, explaining that the fixed stars are on a sphere which rotates around the Earth every day, carrying the other spheres (home to the other celestial bodies) with it. Next comes an explanation of the mathematics Ptolemy will use to calculate the movement of the planets in 'epicycles'. Ptolemy then deals in turn with the motion of the Sun and the Moon, and offers his theory of solar and lunar eclipses. He then discusses the fixed stars, maintaining that their positions relative to one another are unchanging. He gives his own extensive star catalogue at this point in the treatise, containing more than 1,000 stars. The final five books are devoted to his explanation of the movement of the planets.

Ptolemy extracted and improved the tables from the *Almagest* for separate circulation, and wrote an abridged, simpler version for non-experts known as the *Planetary Hypothesis*. Like many astronomers since, Ptolemy also wrote about astrology. Although the two seem strange bedfellows to modern eyes, Ptolemy saw no difficulty: astronomy explains how the celestial bodies move, and astrology explains the effects on humans of those movements. In addition, he wrote on geometry, optics and geography, but his fame rests on the *Almagest*. The complex mathematical model he constructed explained the observed movement of the planets well enough to be taken as the correct explanation for many centuries.

have a sidereal period of one year, there's no very obvious way of choosing. It is worth saying that the planets themselves were not deemed to move: they were fixed to orbs, and the orbs moved. There was, consequently, more than one orb per planet as it had to trace out more than one kind of movement.

With the planetary motions explained there was then no reason to challenge the model and Ptolemy's version of the Earth-centred universe dominated astronomy until the 16th century. So, for 2,000 years, Aristotle's preferred model went virtually unchallenged: the Earth sat immobile at the centre of a succession of concentric transparent spheres that revolved at different speeds around it.

Drifting planets

The Ptolemaic model of the universe explained the pattern of planetary movement and was accepted without question for many centuries. But as observations became more accurate (and the positions of stars relative to the Earth slowly changed over centuries), discrepancies became increasingly obvious.

Muslim astronomers were the first to criticize Ptolemy's model. Abu Sa'id al-Sijzi (951–1020) challenged the notion that the Earth was stationary, suggesting that it revolves on its axis. He even made an astrolabe (see page 75) based on the assumption that it was the Earth which moved, rather than the stars and planets. Aristotle and many subsequent thinkers had rejected this proposal on the basis that if the Earth were moving, in any way, an object thrown or dropped would be left behind as the Earth moved under it, and would therefore not fall straight down to the ground. Others tinkered with

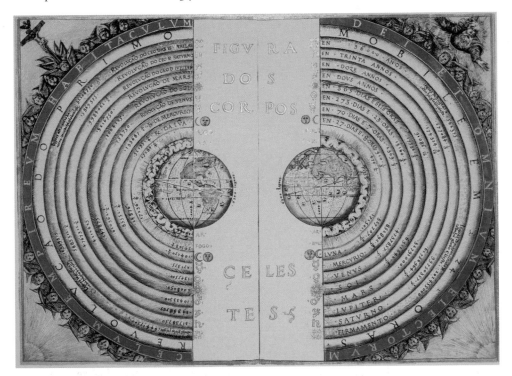

The Ptolemaic model of the universe persisted into the 17th century, by which time angelic hordes had been added outside the 'primum mobile', the sphere which imparts movement to the rest. This region, being outside the physical universe, was the domain of God.

the geometrical model of the planetary motions, but none provided any long-term improvement in accuracy on Ptolemy's original scheme.

A long silence

The works of Aristotle and Ptolemy were lost to Europe for around 600 years, from the 6th to 12th centuries. They were sustained in the Arab world, where they continued to be read and commented on, and were later reintroduced to Europe, often through translations from Arabic. But while the Arab astronomers built instruments and made accurate observations of the night sky, there was a division between astronomy and cosmology. The factual details of orbits and star positions was the business of astronomy; theoretical pondering about the nature of the universe and how it might work was cosmology, and the work of natural scientists and philosophers. The Arabs excelled at astronomy, but their influence on cosmology, outside the realm of religion, lay largely in uncovering flaws in the prevailing Ptolemaic model.

When the works of Aristotle and Ptolemy re-entered the European academies, most time was spent trying to reconcile their scientific deductions with the teachings of the Bible. In particular, Aristotle's claim that the universe is eternal was difficult to reconcile with the Biblical narrative of a time of Creation and an end in Judgment Day.

Prising apart the spheres

As the accuracy of observations increased, it was inevitable that some drastic modification of the Ptolemaic model would be needed. Change began in southern India; Nilakantha Somayaji (1444–1544) developed a system in which the planets Mercury, Venus, Mars, Jupiter and Saturn orbit the Sun. He didn't quite grasp the nettle, though – his small heliocentric system orbited the Earth in its entirety. Most of the Keralan astronomers who followed him accepted the model. The Danish astronomer Tycho Brahe proposed much the same idea in the late 16th century.

The Ptolemaic universe eventually ground to a halt in the 17th century. But the world was not ready to give up its position at the centre of all things without a fight.

The Copernican revolution

In 1543, the Polish astronomer Nicolaus Copernicus reconfigured the universe around the Sun. It's difficult now to appreciate how revolutionary this was. The Ptolemaic model had prevailed for 1,700 years and was fully supported by the Church. Challenging the accepted model was dangerous. The Church was committed to the geocentric model because it upheld the teachings of the Bible: the Earth is special, the haven created by God for humankind, with the rest of the universe serving it. Jostling the Earth out of its central position, making it one of several planets orbiting the Sun, was a serious challenge to this specialness. The Church responded (not immediately, but soon after) by stating its opposition to the heliocentric model and later outlawing it entirely.

Copernicus's model was not entirely original. As we have seen, Aristarchus of Samos had been there before him, and both Indian and Arab astronomers had also proposed it. Indeed, it's possible that Copernicus took at least some mathematical

The Copernican model of the universe put the Sun at the centre and demoted Earth to the same position as the other planets in orbit around it. This image, from 1660, includes the moons of Jupiter, discovered by Galileo Galilei in 1610.

elements from the work of Ibn al-Shatir. But it was Copernicus who changed the world.

Copernicus first presented his heliocentric ideas in the pamphlet *Commentariolus*, which was never printed but circulated in manuscript form between

'We regard it as a certainty that the earth, enclosed between poles, is bounded by a spherical surface. Why then do we still hesitate to grant it the motion appropriate by nature to its form rather than attribute a movement to the entire universe, whose limit is unknown and unknowable? Why should we not admit, with regard to the daily rotation, that the appearance is in the heavens and the reality in the earth? This situation closely resembles what Vergil's Aeneas says: "Forth from the harbour we sail, and the land and the cities slip backward."'

Copernicus, 1543

NICOLAUS COPERNICUS (1473–1543)

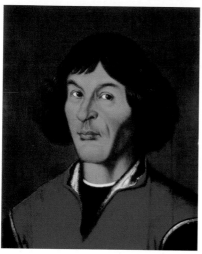

Copernicus was born into a wealthy and politically influential family. A polymath and a polyglot (he spoke German, Latin, Polish, Italian and Greek), he had significant achievements outside astronomy. But it is for astronomy that he is most famous and in which he had most impact. Copernicus first became interested in astronomy while at university in Crakow, Poland, in 1491–4. After graduating, he was appointed canon at Frombork cathedral in Torun, a position he held for the rest of his life. From 1496 to 1503 he left the cathedral to study in Italy, where he became friends with the Italian astronomer Domenico Maria Novara, one of the few people who dared challenge the Ptolemaic model. From 1510, Copernicus lived in cathedral lodgings and practised astronomy in his spare time. He began developing his heliocentric model of the universe around 1508, and in 1513 built a small observatory to aid his studies.

In 1514 he published a pamphlet, *Commentariolus*, outlining his heliocentric model. Copernicus sent the pamphlet to close friends only. Even so, a buzz developed around the ideas over the next few years. When he explained his model fully in 1543 in *De revolutionibus orbium coelestium* ('On the Revolutions of the Heavenly Spheres'), it attracted criticism from other astronomers and the Church. Martin Luther, leader of the Protestant Reformation, also rejected it. The Lutheran minister Andreas Osiander wrote a preface, which was attached to the book without Copernicus's knowledge, saying that the model was only a theory. This was in accordance with an earlier compromise the Church had made with natural philosophy. In 1277, the Church had condemned 227 ideas derived from the teachings of Aristotle, making it an excommunicable offence to hold any of them as true. Philosophers Jean Buridan (c.1295–1363) and Nicole Oresme (c.1320/5–82) engineered a nominalist compromise: science should be deemed to be a working hypothesis that agrees with observed phenomena, but is not taken to constrict what God might be doing with the world. This meant anything could be proposed, as it was recognized that God could have ordered things very differently had He so wished.

De revolutionibus went out with its prologue as though added by Copernicus himself, but at least some readers discovered it had been added by Osiander. Copernicus was already ill and close to death and not in a position to defend his work. He would not live to see the rumpus he had caused.

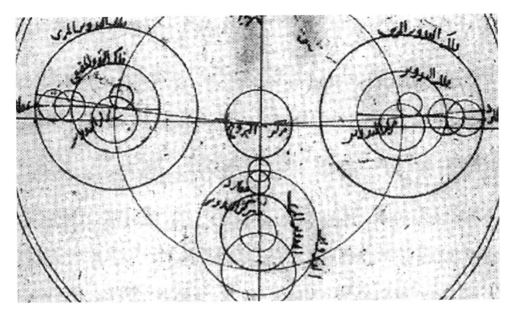

Ibn al-Shatir (1305–75) eliminated the equant and eccentric and added more epicycles to Ptolemy's model to explain the movement of the planets (Mercury in this case). Copernicus might have known of al-Shatir's work; his mathematical model of the movement of Mercury is identical to that of al-Shatir.

1508 and 1514. This pamphlet proposed seven axioms, serving as an announcement of Copernicus's ideas:

1. There is no one centre of all the celestial orbs or spheres.
2. The centre of the Earth is the centre of the lunar sphere – the orbit of the Moon around Earth.
3. The Sun is near the centre of the universe, and all celestial bodies rotate around it.
4. The distance between the Earth and Sun is only a tiny fraction of the stars' distance from the Earth and the Sun.
5. The stars do not move; if they appear to move it is because the Earth itself is moving.
6. Earth orbits the Sun, making it appear that the Sun moves on a yearly cycle.
7. The apparent movement of the planets, with alternating forward and retrograde motion, is an illusion produced by the Earth's movement around the Sun.

It would take Copernicus decades to complete the mathematics and explanations needed to support these statements; his fully articulated theory was published in 1543 in *De revolutionibus orbium coelestium*.

The Church banned *De revolutionibus* in 1616, 73 years after Copernicus's death, and it remained on the list of forbidden books until 1758. Ironically, the Church's reform which introduced the Gregorian calendar in 1582 was based on tables using Copernicus's methods and model.

Changing the world

Outside the Church, *De revolutionibus* was soon influential. Among those who

EPISTEMOLOGICAL RUPTURE

Copernicus's proposal of a heliocentric universe is an example of 'epistemological rupture'. Defined by Gaston Bachelard in 1938, this is the removal of a barrier to scientific advance that has been created by an un-thought-out attitude or belief structure. The idea that the Earth was at the centre of the solar system emerged naturally from observing the Sun move across the sky each day. A scientific model – the Ptolemaic universe – was built around this belief without examining it. The model then became an obstacle to further development. It takes some effort and energy to identify and then overturn such an obstacle. Bachelard proposed that the history of science is a series of such models being developed and then ruptured.

easy ride. Counter arguments included that the Copernican universe contravened the rules of Aristotelian physics, that the predictions it produced were no better than those of the Ptolemaic model, and that it contradicted scripture. Even before the book's publication, Martin Luther, instigator of the Protestant Reformation,

Copernicus set out his model fully in 1543 in his book De revolutionibus orbium coelestium, *one of the most important scientific texts ever published.*

adopted the Copernican system were English astronomers John Dee, Robert Recorde, Thomas Digges and William Gilbert; Michael Mästlin in Germany, and Giambattista Benedetti and Giordano Bruno in Italy. From 1561, the University of Salamanca in Spain allowed students to choose between studying Ptolemy and studying Copernicus.

Nevertheless, the Copernican model did not displace the Ptolemaic immediately. There were many who contested it, even outside the Church. The intellectual elite were still committed to the authority of Classical authors, and a challenge to an idea which had Aristotle's stamp of approval and had held sway for nearly two millennia was never going to have an complained that: 'This fool wishes to reverse the entire science of astronomy; but sacred Scripture tells us that Joshua commanded the Sun to stand still, and not the Earth.'

'Humanity has perhaps never faced a greater challenge, for by [Copernicus's] admission [that the Earth is not at the centre of the universe] how much else did not fall in dust and smoke: a second paradise, a world of innocence, poetry and piety, the witness of the senses, the conviction of a religious and poetic faith. . . . No wonder that men had no stomach for all this, that they ranged themselves in every way against such a doctrine.'

Johann Wolfgang von Goethe, 1810

A compromise position

Among those who did not accept the Copernican model in its entirety was the Danish astronomer Tycho Brahe, last of the great naked-eye astronomers. His own challenge to astronomy, and the root of his own model of the solar system, came from two observations he made in the 1570s.

In 1572, Brahe spotted a new star (see box, page 59) and in 1577, on his way home from a fishing trip, saw a bright comet. His published accounts and explanations offered a significant challenge to astronomy. The two phenomena were at odds with the notion, propounded by Aristotle and

Tycho Brahe's statue in the Botanic Gardens in Copenhagen, Denmark, near the site of his island observatory.

TYCHO BRAHE (1546–1601)

The Danish astronomer Tycho Brahe looms larger than life in the story of astronomy, distinguished particularly by his metal nose and his pet moose. Brahe bought his first astronomical instruments as a student in Germany in the 1560s. In 1566 he lost part of his nose in a duel with another student and wore a metal prosthetic nose, or at least a patch over the gap, for the rest of his life. He returned to Denmark in 1570, where in 1572 he observed the appearance of a new star in Cassiopeia and published a pamphlet about it. The star, now known as SN 1572, was a supernova (an exploding star); its remnants were discovered in the 1960s.

King Frederick II of Denmark gave Brahe funding for an observatory, which he built on the small island of Hven in the Sont near Copenhagen. Called Uraniburg, it became the finest observatory in Europe. At Uraniburg, Brahe built and calibrated new instruments, ran his own printing press and trained young astronomers. His observatory made observations every night, not just at high points of the planets' various orbits as other observers tended to do. It achieved a higher degree of accuracy than any previous observatory: Brahe's observations were accurate to 2 arc minutes (even ½ an arc minute in some cases), while previous observers were generally accurate to around 15 arc minutes. (An arc minute is ⅟₆₀th of a degree, with 360 degrees to a full circle.) Brahe was also the first astronomer to make corrections for atmospheric refraction – the effect of Earth's atmosphere in slightly distorting the view of the stars.

A dispute with the new king, Christian IV, led to Brahe leaving Denmark for good in 1597. He settled in Prague in 1599, but lived only another two years before his death. During that time, another great astronomer, Johannes Kepler, worked with him. Kepler later discerned the elliptical orbit of planets from Brahe's observations.

And the moose? Brahe had a pet moose that attended feasts and banquets with him. One day it drank too much beer, fell down the stairs of the castle they were visiting and died. Brahe's own death was also dramatic. Refusing to leave a feast to urinate, he either damaged his bladder or contracted a urinary infection which killed him eleven days later.

the Church, that the heavens are eternally unchanging. The supernova and comet were clearly beyond the Moon, probably among the fixed stars.

Brahe demonstrated that the comet was definitely not within Earth's atmosphere. He did this by comparing his observations of the comet near Copenhagen with observations made by Tadeáš Hájek in Prague at the same time. The comet was in substantially the same position when viewed from both locations, while the position of the Moon was significantly different, so Brahe deduced that the comet must be further away than the Moon. If Ptolemy had been right about the fixed shells of the celestial spheres, the comet must move between them – but this was clearly impossible. This part of the Aristotelian–Ptolemaic model could not be sustained, and the idea of physical celestial spheres disappeared during the period 1575–1625.

However, Brahe was not ready to renounce the Ptolemaic model entirely. Instead, he developed a sort of halfway house, known as the Tychonic model, which had all the planets except Earth orbiting the Sun, then the Sun and Moon both orbiting Earth. The sphere of the fixed stars also went around the Earth. Tycho Brahe was committed to the idea of a stationary Earth, complaining that the Copernican model 'ascribes to the Earth, that hulking, lazy body, unfit for motion, a motion as quick as that of the aethereal torches', which he found implausible. The newly discovered comet he put in orbit around the Sun, between Venus and Mars. His model was popular in the early 17th century among people disenchanted with the Ptolemaic universe and its necessary fudges, but unwilling to embrace the Copernican heliocentric model.

Round and round?

One reason that astronomers did not all flock to adopt Copernicus's model is that the predictions it gave of planetary movement were no more accurate than those of the Ptolemaic model; it still required equants and epicycles to match what was seen in the heavens. That was all swept away by the pupil, assistant and successor of Tycho Brahe, Johannes Kepler (see box, page 61).

Copernicus's biggest mistake was to assume that the planets were in circular

'On the 11th day of November in the evening after sunset, I was contemplating the stars in a clear sky. I noticed that a new and unusual star, surpassing the other stars in brilliancy, was shining almost directly above my head; and since I had, from boyhood, known all the stars of the heavens perfectly, it was quite evident to me that there had never been any star in that place of the sky, even the smallest, to say nothing of a star so conspicuous and bright as this. . . . A miracle indeed, one that has never been previously seen before our time, in any age since the beginning of the world.'

Tycho Brahe, 1572

The comet observed by Tycho Brahe in 1577, depicted by Jiri Daschitzky. The comet is visible in front of the thick cloud, reflecting the belief current at the time that comets were within Earth's atmosphere.

BRAHE'S COMET

The comet that Brahe saw, now officially designated C/1577 V1, was one of the five brightest comets of recorded history. It is a long-period comet, not expected to return for thousands of years, and currently located more than 300 AU from the Sun. (One AU, or Astronomical Unit, is the distance between the Earth and the Sun.)

orbits around the Sun. Working from Brahe's meticulous and comprehensive data from observations, Kepler deduced in 1605 that the planetary orbits are elliptical, not circular. He was not the first to suggest this. Both the Indian astronomer Aryabhata (476–550) and the Muslim astronomer Abu Ma'shar al-Balkhi (787–886) had already described the Earth following an elliptical orbit around the Sun. But neither had sufficient influence, nor mathematical backup, for the idea to be widely accepted.

JOHANNES KEPLER (1571–1630)

Johannes Kepler was born to a mercenary soldier and the daughter of an inn-keeper in Swabia (southwest Germany). When Kepler was five, his father died at war and the boy grew up in his grandfather's inn. He went to university at Tübingen, where he was taught by the great astronomer and mathematician Michael Mästlin (1550–1631). Kepler learned the official Ptolemaic astronomy but, as he was a favoured pupil, Mästlin also introduced him to the Copernican model. Kepler was convinced of the validity of the model immediately.

Kepler was both devoutly religious and a talented mathematician. For him, there was no difficulty in reconciling the two: he considered that God had made the universe according to a mathematical plan and humankind could understand the plan through mathematics. The Church was not as well persuaded, however, and he was excommunicated in 1612 for unorthodox beliefs. This was never reversed and caused him much pain. Nevertheless his mathematics brought him success as an astronomer; his was the first accurate mathematical model of the solar system.

Kepler's first cosmological model, published in *Mysterium cosmographicum* ('Mysteries of the Cosmos') in 1596, was esoteric. His calculated figures did not quite match observations; Mästlin hoped for a closer match if Kepler could gain better data to work with. To this end, he sent a copy of the book to Tycho Brahe. Brahe had also just written to Mästlin as he wanted a mathematical assistant. Kepler got the job, perhaps one of the most fortunate appointments in the history of science.

Brahe was old and deeply engrossed in what he called his 'war with Mars', trying to work out the nature of the planet's orbit. He made many detailed observations and records of the path of Mars. Since orbits were assumed to be circular, it was common to make just a few observations and then plot the orbit from the known positions and radius. Brahe did not win the war with Mars before his death in 1601; Kepler inherited all his data and the problem with Mars as well as his post as Imperial Mathematician. With Brahe's exhaustive data, and his own mathematical skill and dogged perseverance, Kepler eventually calculated that Mars has an elliptical orbit, with the sun at one focus of the ellipse. He soon found this could be extended to the other planets, and published his findings in 1609 in *Astronomia nova* ('New Astronomy'), with the first two of his planetary laws (see page 64).

- The law of ellipses: The planets trace elliptical orbits around the Sun, with the centre of the Sun being one focus of the ellipse.
- The law of equal areas: An imaginary line drawn between the centre of the planet and the centre of the Sun will sweep out equal areas in equal intervals of time. (Although the speed of the planet varies, being faster when it is closer to the Sun, the area covered is always the same in any period.)

The final law was published much later, in 1619. It emerged from further work Kepler carried out using Brahe's data, and was called the law of harmonies:

- The ratio of the squares of the periods of any two planets is equal to the ratio of the cubes of their average distances from the Sun.

Planet	Period (s)	Average Distance(m)	T^2/R^3 (s^2/m^3)
Earth	3.156×10^7 s	1.4957×10^{11}	2.977×10^{-19}
Mars	5.93×10^7 s	2.278×10^{11}	2.975×10^{-19}

Although his spheres have faded into obscurity, Kepler's laws of planetary motion were a crucial development.

The tables he compiled from the data Brahe had begun to collect were important not only in being the most accurate ever published, but in another, unexpected way. Up until this point, all such tables that set out the motion of the planets in relation to the stars had been based on the Ptolemaic geocentric model of the universe.

However accurate the Ptolemaic tables were when first published, they did not match the movements of the celestial bodies for long and soon had to be revised. Kepler's tables were based on the Copernican model and were found to be accurate not just for years, but for decades. Kepler would have been gratified to learn this, but he died only three years after their publication. That his tables remained accurate was further – critical – support for the Copernican model and helped to accelerate its adoption by astronomers.

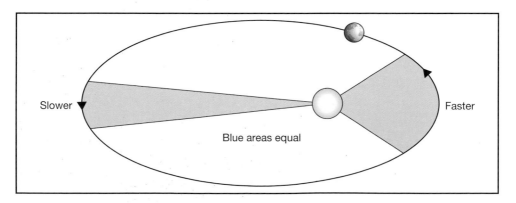

Kepler's second law shows that a planet moving in an ellipse around the Sun will cover equal areas in equal amounts of time. The planet moves faster when it is nearer the Sun.

THE RUDOLPHINE TABLES

The object of all Brahe's observations, and a task which fell to Kepler on Brahe's death, was to compile astronomical tables. These showed the position of the planets with relation to the fixed stars on different dates during the year. The tables most commonly used at the time were the Alphonsine tables, commissioned by Alfonso X of Castile. Produced in the 13th century and updated intermittently thereafter, they were based on the Ptolemaic model and were not very accurate. Brahe's intention had been to replace them with comprehensive and accurate tables and a new star catalogue. The tables were finally completed by Kepler,

using Brahe's data as well as his own, in 1623, and published in 1627. The tables were named after the Holy Roman Emperor, Rudolf II, even though he had died by the time they were published.

During the course of his work, Kepler discovered the logarithmic tables published by the Scots mathematician John Napier in 1614. Although Kepler's tables were assumed by many to be more accurate than others, given that they rested upon Brahe's data, in practice the logarithms were tedious and cumbersome to use. Kepler also included Brahe's catalogue of 1,000 fixed stars, and instructive examples for computing planetary positions.

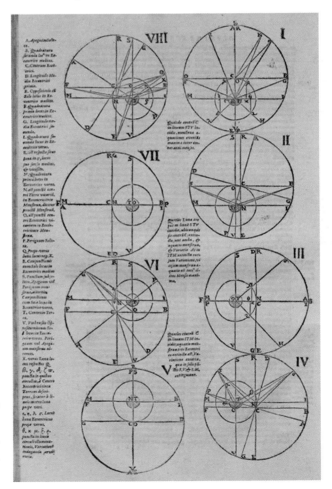

The threshold of the modern age

In one of those remarkable moments of serendipity, the same year that Brahe published his laws of planetary motion, the Italian astronomer and mathematician Galileo Galilei first peered at the heavens through a telescope of his own making (see page 79). This was the great watershed in the history of astronomy. The shift from naked-eye observation to the use of the telescope in 1609 revolutionized and redefined the science. It is remarkable that that the true nature of the solar system and its relation to

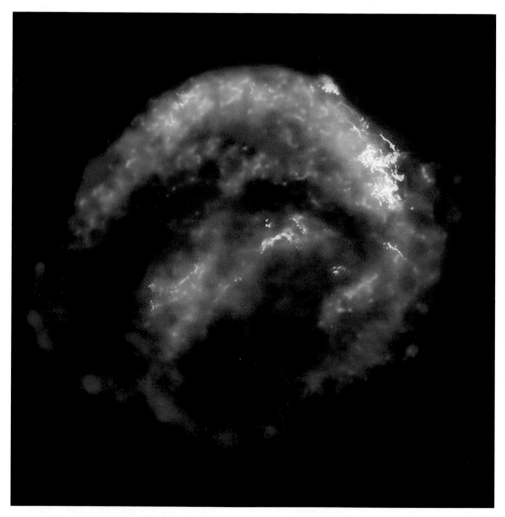

Four hundred years ago, sky watchers, including Johannes Kepler, saw a new star in the western sky, rivalling the brilliance of the nearby planets. Today, astronomers using NASA's three Great Observatories are unravelling the mysteries of the expanding remains of Kepler's supernova.

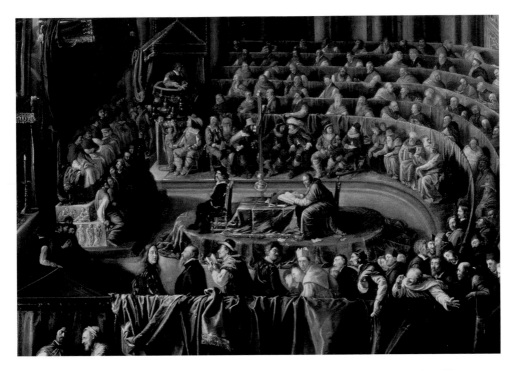

At Galileo's trial in 1633, he was found guilty of teaching that the Earth moves around the Sun. He was excommunicated and spent the rest of his life under house arrest.

the fixed stars were already known, even if not universally accepted, before the advent of the telescope. Cosmology still had a long way to go, but the crucial recognition that the planets orbit the Sun in elliptical paths and that mathematics – not deities – hold the key to their motion was already in place. The age of the telescope would bring far greater knowledge about what there is in the universe. Mathematics and mechanics would begin to explain how it all works.

It is worth remembering, though, that we are only 400 years on from Kepler's revelation of the elliptical orbit and still less than 500 years on from Copernicus's *De revolutionibus*. The Aristotelian–Ptolemaic model reigned for four times as long. And although the scientific community was increasingly swayed by the Copernican model, the Catholic Church became increasingly hostile towards it over the course of the 17th century. *De revolutionibus* was banned in 1613, and Galileo was excommunicated in 1633 for teaching the Copernican model, his own book also banned (see page 81). There was a long way to go.

> '*I hold that the Sun is located at the centre of the revolutions of the heavenly orbs and does not change place, and that the Earth rotates on itself and moves around it.*'
>
> Galileo Galilei, 1616

CHAPTER 3

Tools of
THE TRADE

'Earth is the cradle of humanity, but one cannot remain in the cradle forever.'

Konstantin Tsiolkovski,
Russian rocket scientist,
1895

Astronomy is not really amenable to the forms of experimentation available to the other sciences. We can't set the planets on an alternative course to discover whether gravity works better in a different way, nor can we form a new planet or star to watch its development. Observation and measurement remain astronomers' principal methods of investigation. We can then construct mathematical models and compare them with the observational data to see how well they match. The instruments that have been used for astronomy have consequently been tools for observing and measuring.

Astrolabes have been in use for more than 2,000 years to locate stars and planets.

Line of sight

The first people to look at the night sky and record the paths of the planets and stars had nothing but their own eyes to help them. But over time people began constructing astronomical tools, both large and small, to track the movements of celestial bodies.

First tools

The earliest known astronomical tool is the gnomon, the part of a sundial used to cast a shadow. In its simplest form, it is just a vertical rod or stick attached to a plate with appropriate markings. It was used by astronomers in Ancient Babylon and introduced from there to Ancient Greece by Anaximander in the 7th century BC. A gnomon was (and is still) used to find the declination of the Sun. (The declination of a celestial body is the number of degrees that it is north or south of the celestial equator.) The gnomon is calibrated for noon at the observer's position (latitude); that is the point when the shadow of the gnomon is shortest.

Another very simple tool is the dioptra. This consists of a sighting tube, or a rod with sights at both ends, mounted on a circular stand that allows it to be turned precisely. It emerged in Greece in the 3rd century BC. If the edge of the stand is marked with degrees, the dioptra can be used to measure the angles of the positions of stars. These are very simple tools – which makes the sophistication of the next oldest instrument all the more astonishing.

The gnomon of a sundial represents the simplest and earliest astronomical tool.

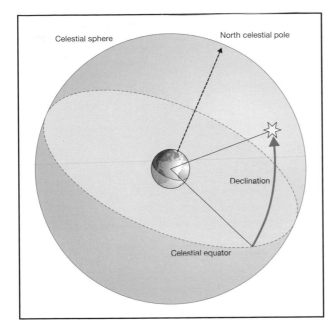

Celestial sphere

North celestial pole

Declination

Celestial equator

The North celestial pole is directly above Earth's North pole; the celestial equator follows Earth's equator at a distance; the declination of the Sun or a star is its angle from the celestial equator towards the celestial pole.

even been claimed by some to be an 'out of time' object – an item so far out of its correct chronological/archaeological setting that time travel or alien intervention have been suggested to account for it.

The explanation turned out to be less fanciful, though still extraordinary. In 2006, British and Greek scientists working together discovered that the mechanism is a machine for calculating the exact positions of the Sun, Moon and planets in the sky.

A celestial computer

In 1900, sponge-divers working in the Aegean Sea off the coast of the Greek island Antikythera discovered a shipwreck dating from the 1st century BC. Among the many artefacts on the wreck was one that has been named the Antikythera Mechanism. For decades, the function of this corroded lump of metal remained obscure. Despite the corrosion, it was clear that it once had a complex system of gears and was a finely engineered piece of machinery. Indeed, so out of place did it seem in an ancient shipwreck that it has

The Antikythera Mechanism, now in the National Museum in Athens.

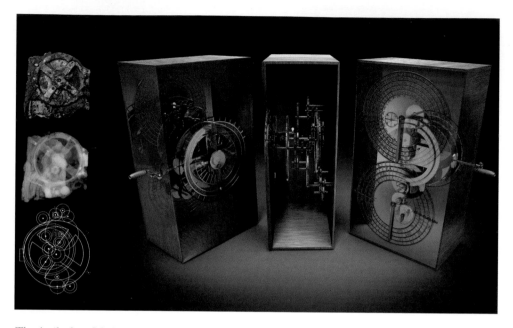

The Antikythera Mechanism re-created to show its workings, with (left, top to bottom) a photo, an X-ray, and a computer model of the relic.

It had a system of 30 finely graded gears that were turned using a handle on the side. It would have been mounted in a wooden case (fragments of the wood remain), and was about the size of a mantel clock. Like a clock, it had a large dial on the front. Seven hands or pointers moved to reproduce the movement of the Sun, Moon and five known planets. A small silver and black ball rotated to show the phases of the Moon, even reproducing the changes in the Moon's speed over a nine-year cycle. Inscriptions reveal that the pointers (now lost) carried graphic representations – the Sun had a fiery ball, and Mars a red ball. The movement replicated the effects of the epicycles (as described by Ptolemy). There were two additional dials on the back of the mechanism. One worked as a calendar, and the other was used to predict solar and lunar eclipses. The mechanism probably dates from the 1st or 2nd century BC, and is possibly from Rhodes.

Modelling the celestial globe

Nothing remotely comparable would be made again for 1,600 years (when Blaise Pascal made the first mechanical calculator). The Antikythera Mechanism remains a unique find; if other astronomers had similar instruments, no record or evidence of them survives. Instead, two other types of mechanical model of the universe were used to depict and explore the night sky, one that worked in three dimensions and one that worked in two: the armillary sphere and the astrolabe. Often ornate and carefully crafted, they lacked the complexity and range of the Mechanism, but were created and used over many centuries.

Armillary sphere

According to the Greek astronomer and mathematician Hipparchus (190–120BC), the armillary sphere was invented by Eratosthenes (276–194BC). It consists of a central sphere, representing the Earth, surrounded by a series of bands which form a sort of skeletal sphere. This sphere represents important circles in measuring angles in the sky. It was set up for the observer's location by correctly lining up the band representing the meridian (an imaginary line that runs perpendicular to the horizon) on a north–south axis and then finding a star whose position on the ecliptic was known. The ecliptic coordinate positions of other celestial bodies could then be found.

It's possible that a simple armillary sphere was in use in China a little earlier (during the 4th century BC) but this is contested. It is known, though, that a simple form with a single ring was in use in the 1st century BC in China. A second, equatorial, ring was added by Geng Shou-chang in 52BC and the armillary was further improved by the great astronomer Zhang Heng. His final armillary was bronze, 5m (16½ft) across, and powered

by a water clock and gears so that it rotated slowly. It had a central tube that was used to line up the stars and planets.

In Europe, too, the earliest armillary spheres had only one or two rings. Eratosthenes probably had an armillary

An armillary sphere has a model of the Earth at the centre and bands representing important circles such as the ecliptic, celestial equator and meridian.

ZHANG HENG (AD78–139)

Zhang Heng was born in the Han dynasty to a distinguished family. Trained first as a writer, he turned his attention to science at around the age of 30. He was an accomplished astronomer, mathematician and engineer, inventing the first odometer (to measure distance) and the first seismoscope (to respond to seismic movement and warn of impending earthquakes). He considered the universe to have originated from chaos. He showed that the Moon does not shine with its own independent light, but with reflected sunlight, and he was aware that space might be infinite. He also corrected the Chinese calendar in AD123, bringing it into line with the seasons. Zhang remapped over 2,000 stars and added a geared device to his giant armillary sphere to show the waxing and waning of the moon. He divided the stars into 124 constellations and named the brightest 320 stars. He was aware his map was not exhaustive, saying that there are 11,520 less bright stars.

A cutaway of Zhang Heng's seismoscope shows how it works: a pendulum linked to a rod connecting the dragons' mouths is displaced by an earthquake or tremor, causing a dragon mouth to open and drop a ball into the mouth of a toad beneath. The direction of the quake is indicated by which toad receives the ball.

sphere in the 3rd century BC with a band representing the plane of the equator, crossed by one representing the meridian. Armillary spheres became progressively more complex, adding more bands that represented other important features, including the ecliptic. Ptolemy, describing his armillary sphere in the *Almagest* in the 2nd century AD, talked of six rings. His is the first full description of an armillary sphere. Armillary spheres of increasing complexity continued to be used well into the 16th century; some were beautifully ornate objects.

Astrolabe

While an armillary sphere is three dimensional, an astrolabe is a two dimensional map of the heavens. It uses movable discs and fretwork to help the observer locate objects visible from their location at the appropriate time. Inevitably, projecting the hemispherical vault of the sky onto a flat surface involves distortion. Some recognition of this, perhaps, is evident in the Arab legend that the first astrolabe was made accidentally by Ptolemy. According to the story, Ptolemy was riding on a donkey while using his armillary sphere (which would surely count as riding without due care and attention) when he dropped it and the donkey trod on it, squashing his three-dimensional sphere into a flat disc. It is not certain quite when the astrolabe first appeared. Its invention has been credited to Hypatia, who lived in Alexandria in the 4th century BC, and also to Apollonius of Perga in the 2nd or 3rd century BC in Greece. The oldest surviving astrolabe is Islamic and dates from 927 or

Ptolemy with his armillary sphere, as imagined by an artist in 1476.

928, so it is impossible to know exactly what earlier astrolabes were like.

An astrolabe has a circular back plate called the 'mater' with degrees marked around the edge. A disc, or plate, engraved with circles of azimuth (celestial latitude) and altitude is fastened onto the mater. Above this lies a metal fretwork called the 'rete', marked with the positions of bright stars. This looks like an ornate and rather spidery filigree as the stars are scattered all over the place but the plate behind must remain visible, so thin tendrils of metal extend out into space as pointers to show the positions of the stars.

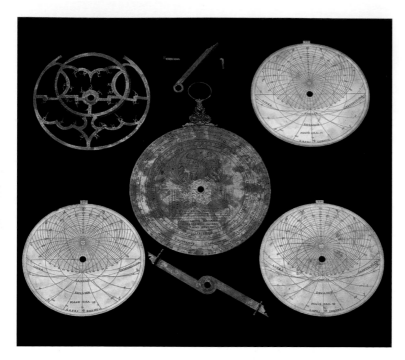

The components of an astrolabe: the mater is in the centre, the rete at top left, and the other plates are arranged around the mater.

The user finds the altitude of a bright star using a sighting device, either integral to the astrolabe or separate, then turns the rete until that star is aligned with the correct line of altitude on the plate. At this point, all the other stars on the rete are shown in their proper positions for that latitude and time and the user can enjoy a happy night of star-spotting.

In the service of Allah

Arab astronomers were driven by the needs of Islam for accurate time-keeping and location-finding. The five prayer sessions of each day had to be matched to specific positions of the Sun, and the direction of Mecca had to be known from any location in the Islamic world so that the faithful could pray while facing the holy city. The Islamic calendar is based on the lunar cycle,

so measuring and predicting the Moon's movement was also important. There was plenty of work for astronomers. The

CHAUCER'S TREATISE ON THE ASTROLABE

The English author Geoffrey Chaucer (c.1343–1400), most famous for *The Canterbury Tales*, wrote a *Treatise on the Astrolabe* which sets out to explain to a young child how to make and use an astrolabe. It is the first technical manual written in English. Its popularity (it survives in many manuscript copies) suggests a keen interest in astronomy among the middle- and upper-classes, who could afford to have such manuscripts copied.

STARTING FROM PTOLEMY

Although the Arabs agreed with Ptolemy that the movement of celestial bodies is the result of natural laws, they disagreed with him on some details. Al-Battani (858–929) was among the first to find fault with Ptolemy. His *Klitabal-Zij* pointed out some of the errors Ptolemy had made in reporting planetary motion. He also discovered that the solar apogee (aphelion, the point at which Earth is furthest from the Sun) slowly moves, that the ecliptic is tilted, and that the Greeks had miscalculated the rate of precession.

The errors and inconsistencies that the astronomers of the 9th century found as they investigated Ptolemy's text and compared it with their own observations led to a more thorough and critical approach. They made ever more precise measurements over the coming centuries, but their refinements did not make their understanding of the Earth's place in the heavens any more accurate than Ptolemy's version.

Astronomers working in the observatory of Taqi ad-Din, Istanbul, completed in 1577. It was destroyed in 1580 in response to objections to prognostication (prophesying future events).

requirements of astronomy also drove progress in mathematics, particularly trigonometry. Arab astronomers developed giant instruments to make more accurate measurements, and built well-equipped observatories, staffed by many expert astronomers.

Giant instruments

In Samarkand, Uzbekistan, a trench runs along the floor of what used to be a great observatory. The observatory belonged to the astronomer Ulugh Beg (1394–1449) and the trench, edged with marble slabs, is all that remains of his giant astronomical quadrant, an instrument for determining the angle of elevation of celestial bodies. With a radius of around 40m (131ft), it is the largest known. In its heyday, Beg's quadrant traced out a quarter of a circle, running through the trench in the floor and up the curved wall. It was marked in degrees, minutes and seconds, essentially forming a giant protractor. High on the opposite wall was a small slit window, right at the centre of the circle defined by the quadrant. Astronomers working in the observatory used the giant quadrant with its sighting slit to record the positions of stars or the Sun. Using it, Beg and his associates al-Kashi and Ali Qushji compiled a remarkable catalogue of 992 stars, called *Zij-i Sultani*, which superseded that of Ptolemy in 1437.

A quadrant provides a 90-degree span and a sextant covers 60 degrees. Confusingly, quadrants are often classed with and even called 'sextants'. The first mural sextant (one built into walls and the floor like that at Samarkand) was made by Abu-Mahmud al-Khujandi in 994. Later framed sextants were smaller but could be moved around.

Beg's giant quadrant had the advantage that, being so large, it could manage greater resolution and so was more accurate than any previous instrument. His observatory calculated the length of a year to be 365 days, 5 hours, 49 minutes and 15 seconds, just 25 seconds out. His astronomers' calculations of the movements of Saturn, Jupiter, Mars and Venus differ from modern values by only 2–5 degrees.

Astrolabes for stars and planets

Arab astronomers copied the astrolabe from the Greeks and improved it, using it to find the direction of Mecca, the date for the beginning of Ramadan and the hours for daily prayers. (Astrolabes could be marked with times around the rim; after finding a reference star and lining up the mechanism, the time could be read.) The earliest known Arab astrolabe dates from 927–8. Al-Zarqali (1029–87) made one that did not depend on location, so could be used anywhere. It became known in Europe by the name 'Saphaea'.

Al-Zarqali also made an equatorium, a device somewhat similar to an astrolabe, but used to find the positions of the Sun, Moon and planets rather than the stars. The Greek philosopher Proclus (AD412–85) also described such a device and how to make one. This, too, continued to be made in later Europe but as it was more difficult to use and less useful than an astrolabe (it didn't deal with the stars), it was less common and few examples survive. It seems to have been used predominantly for astrological purposes.

A new way of looking

If there is a single defining moment that marks the start of modern astronomy, it is the first time that Galilei Galileo looked

OPENING UP THE HEAVENS

Although Arab scientists made great advances in optics and lens-making, and Arab astronomers made astonishingly accurate measurements and observations of the night sky, they never put two and two together to make a telescope. That step was taken centuries later, in Europe.

THE AIR GETS IN THE WAY

Ibn al-Haytham (965–1039) measured the thickness of Earth's atmosphere and calculated the effect it had on astronomical observations. Atmospheric interference also troubled the later astronomers who used telescopes, and was only finally overcome in the 20th century with the development of telescopes sited outside Earth's atmosphere, in space.

through his newly made telescope and saw the features of the Moon, the moons of Jupiter and the Milky Way resolved into a band of stars. That moment, when he must have recognized that the Earth is not the only world in the universe, redefined what it is to be human.

In 1609, Galileo received a letter from his friend, the mathematician Paolo Sarpi, telling him about the invention of the telescope. Sarpi had seen one displayed in Venice. Galileo set about making his own telescope, quickly constructing a device that made objects appear 'one-third of the distance off and nine times larger than when they are seen with the natural eye alone'. His telescope comprised two lenses fitted inside a lead tube. Both lenses were flat on one side, but one was concave and the other

In his Aequatorium astronomicum *(1521), Johannes Schöner provided templates for the reader to cut out and make an equatorium for the planet Saturn.*

ÆQVATORIVM SATVRNI.

¶ Deferens Saturni mouetur per unum diem duobus minutis fere, & in anno uno 12 gradibus, 13 minutis, & 35 secundis iuxta eius medium motum, peragit signiferum

GALILEI GALILEO (1564–1642)

Galilei Galileo was one of the greatest intellects in an age of great intellects. Encouraged by his father to train in medicine, the young Galileo went to university in Pisa but spent a lot of his time attending classes in mathematics, which he found more rewarding. He eventually left Pisa without a medical degree and instead taught mathematics for many years.

Already interested in astronomy, Galileo was obliged to teach the conventional geocentric Ptolemaic model, but was personally convinced by the Copernican model. He wrote to Kepler saying as much in 1598. In 1604, he gave three public lectures in which he argued that the new star seen that year, now known as Kepler's supernova (see page 66), lay beyond the realm of the planets and therefore demonstrated that Aristotle was wrong about the unchanging nature of the heavens. Otherwise, though, he remained diplomatically discreet about his Copernican views.

Galileo's work in mathematics tended towards mechanics. He formulated the law of falling bodies and determined that the path of a projectile is parabolic. Then, in May 1609, he heard about the invention of the telescope and began making his own. He first saw the military and commercial uses for ships at sea, but then realized he could use it to view the heavens. So began his career as the first astronomer to use a telescope.

In a move of brilliant cunning, Galileo gave the rights to make telescopes to the Venetian state in exchange for an increase in his salary. This was a touch dishonest: he had not invented the telescope, had no rights to give away, and no restriction on the manufacture of telescopes was remotely enforceable. The advantage did not last long, but it lasted long enough. By the time the state froze his salary, in 1610, he had published a short book on his findings, *Sidereus nuncius* ('The Starry Messenger') and impressed Cosimo de' Medici in Florence, who appointed him official 'Mathematician and Philosopher'. The book caused a sensation and Galileo was fêted as a celebrity. He continued his observations, finding (but not correctly identifying) the rings of Saturn and discovering that Venus, like the Moon, has phases. But trouble was brewing.

Galileo's opponents passed one of his letters to the Inquisition, the Church's tribunal charged with suppressing heresy. In it, Galileo had said of the perceived conflict between the Copernican system and the teachings of the Bible that the Bible must be interpreted in the light of scientific findings. On this occasion, the Church found nothing to object to as they considered Copernicus's theory to be a convenient mathematical model rather than a statement of fact (see page 54). But another letter was not so calmly received. In 1616, Galileo wrote to the Grand Duchess of

Lorraine arguing that Copernican theory was literally true, and that the Bible gave a non-literal account. This inverted the nominalist position negotiated by Buridan in the 13th century and represented a direct challenge to Biblical authority. The Church examined Copernican theory and condemned it. Copernicus's book, *De revolutionibus*, was banned and Galileo was forbidden to hold or teach the theory.

Galileo expected things to get easier in 1623 when Urban VIII, a pope sympathetic to his views, was elected. He began writing his *Dialogo sopra i due massimi sistemi del mondo* ('Dialogue Concerning the Two Chief World Systems'), which sets out the Copernican and Ptolemaic systems and comes down in favour of the Copernican. When Galileo finally published it, in 1632, the Church was not impressed. He had been expected to produce a purely theoretical treatise that gave equal weight to both arguments. Galileo was summoned to Rome, tried, and found guilty of heresy and contravening the ban of 1616. He was sentenced to life imprisonment, which took the form of permanent house arrest. He ended his days in ill health, blind and deprived of the company of his dearest daughter, who died in 1634.

In 1992, Pope John Paul II admitted that the Church had made errors in the case of Galileo and pronounced the case closed, but he did not overturn the conviction for heresy.

convex on the other side. The concave lens was the eyepiece. These early telescopes offered eight-fold magnification, but Galileo soon improved his design to provide 20x magnification.

It has been said that Galileo made more world-changing discoveries in the two months of December 1609 and January 1610 than anyone else has made before or since. Although Copernicus and Kepler had already written of the heliocentric model of the solar system, they wrote in Latin, whereas Galileo wrote in Italian, making his revolutionary findings accessible to the wider public in Italy.

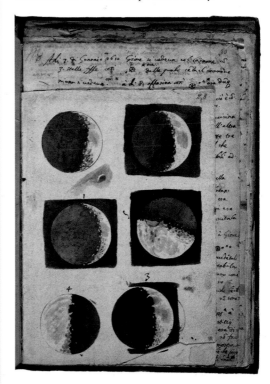

Galileo's watercolours, produced in November– December 1609, are the first realistic images of the Moon ever made.

> *'About ten months ago a report reached my ears that a certain Fleming had constructed a spyglass by means of which visible objects, though very distant from the eye of the observer, were distinctly seen as if nearby.'*
>
> Galileo, 1610

Improving a good thing

In 1611, Kepler described an improved telescope, swapping Galileo's concave lens for a convex lens. This reduced spherical aberration (the blurring caused by any lens curved like a sphere), giving a less distorted image, but the image would be upside down (see box opposite). Although he described the advantages, Kepler did not, as far as we know, make such a telescope. The physicist and astronomer Christoph Scheiner was the first person to do so; he described such an instrument in 1630. Even so, Keplerian telescopes were not widely used until the middle of the 17th century. The first to take considerable advantage of the design was Christiaan Huygens.

Bigger is not always better

Galileo's telescopes were only 120cm (4ft) at their longest point, but a Polish brewer-turned-astronomer, Johannes Hevelius (1611–87), extended the principle. He knew that the flatter the objective lens, the clearer the image – but this also made for a longer focal length. In 1647 he first made a telescope 3.6m (12ft) long that gave 50x magnification. Inspired by its success, he set out to make ever longer telescopes. His largest was 45m (150ft) long. Unfortunately, it was practically unusable. It was too long for a heavy (and expensive) metal tube, so he

Chromatic aberration is the result of light rays of different wavelengths being refracted differently by the same lens. The focus for blue rays is closest to the (convex) lens, and the focus for red rays furthest from it, producing a blurred image.

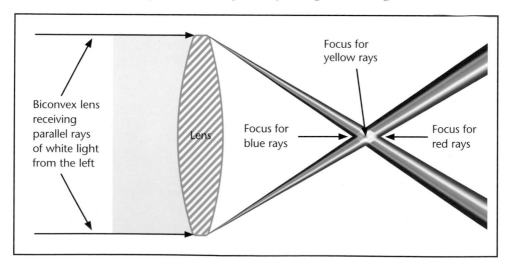

Biconvex lens receiving parallel rays of white light from the left

Lens

Focus for blue rays

Focus for yellow rays

Focus for red rays

FAR-SIGHTED

There are two basic types of optical telescope: refracting and reflecting. The huge, sophisticated optical telescopes of modern observatories still use these same principles at their core.

Galileo's telescope was a refracting telescope. Light enters the tube of the telescope and is focused by a lens. The effect is to gather and concentrate light, so more of it reaches the eye; this enables the observer to see more detail. There are two lenses. The objective lens at the distant end of the telescope is convex and focuses the light. The eyepiece lens, near the observer's eye, is concave and 'straightens out' the converging beams of light again before they enter the eye.

Kepler's improvement put a second convex lens at the eyepiece. The focal point of the objective lens was then in front of the second lens, with the image refocused by the second lens. Because the rays of light have crossed over between the two lenses, the image seen is inverted.

A reflecting telescope uses a convex mirror in place of the objective lens. The big difference is that this time the light enters the telescope from the opposite end (near the observer's eye). A concave (primary) mirror focuses the light on to a second flat (secondary) mirror that is tilted to reflect it towards the eyepiece, where a convex lens straightens it out before it reaches the observer's eye.

Because light of different wavelengths (colours) is refracted differently but reflected equally, a reflecting telescope gives a more precise image with less interference than a refracting telescope.

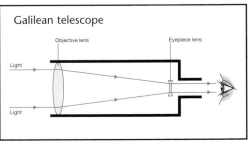

Galilean telescope

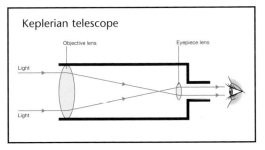

Keplerian telescope

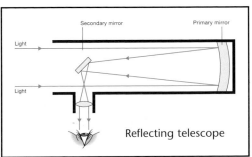

Reflecting telescope

fixed the lenses into a long wooden trough and suspended the whole thing from a pole 27m (89ft) tall. It was controlled – just – by a system of ropes. But it swayed in the slightest breeze and both the wood and the ropes stretched and shrank with changing humidity and temperature, so needed constant readjustment. Occasionally, it collapsed completely. In addition, as it was to be used at night (obviously), all these adjustments had to be made in the dark. It's hardly surprising that it was rarely used.

Another development of the late 17th century was the aerial telescope. This was mounted on a ball-joint on a tall structure, such as a tree or a pole, and connected to the eyepiece by a string or rod. The observer held the eyepiece and used the string or rod to move the telescope.

On reflection

For really significant improvements, a new type of telescope was needed – the reflecting telescope. English mathematician and polymath Isaac Newton (1642–1727), demonstrated a reflecting telescope to the Royal Society in 1672. Newton had successfully made a convex spherical mirror from copper-tin alloy and used it to construct a telescope with 40x magnification. It worked

well enough, but other people had difficulty making the mirrors. Consequently, not much progress was made with reflecting telescopes until 1717 when John Hadley made one with a parabolic, rather than spherical, curve to the mirror. He achieved around 200x magnification and also produced a new type of mount, called the 'alt-az' (altitude-azimuth, or altazimuth) mount, which the astronomer used to move the telescope both horizontally and vertically at the same time.

Late in the 18th century, the German-English astronomer William Herschel (1738–1822) began making telescopes. He started with refractors, but switched to reflectors to avoid the difficulties presented by the long tubes. This was ironic, as his largest reflecting telescope was massive. Although at 12m (40ft) long it was shorter than many previous refractors, the mount took up an enormous amount of space. To use it, Herschel sat, perilously, high up

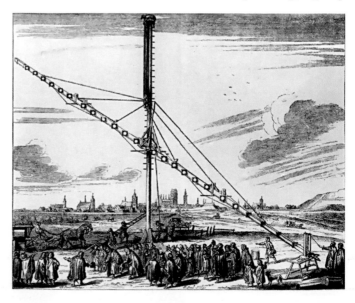

Hevelius's giant telescope looked dramatic, but was not very user-friendly.

Herschel's 12m (40ft) telescope was cumbersome and difficult to use; he generally preferred his smaller model.

in the structure built around his telescope. His first reflector was a modest 2.1m (7ft), but he discovered the planet Uranus with it, and became famous overnight. Using the £200 salary awarded to him by King George III for this discovery, he turned to astronomy full time and built larger telescopes, 6m (20ft) and 12m (40ft) long. He spent many cold nights perched on a platform beneath his vast telescope. In reality, the 12m telescope was less useful and therefore less used than the 6m version, which remained his standard working telescope for most nights. The 12m telescope's mirror needed frequent re-polishing and its tube tended to bend; Herschel did not use it after 1815.

Back to refraction

Refracting telescopes became popular again in the 1750s when the English optician John Dolland developed an achromatic lens that eliminated chromatic aberration. It worked by combining a concave lens and a convex lens with different densities, the second correcting the aberration produced by the first. Still, it was not possible to produce sufficiently high-quality glass to make large lenses. That problem was solved by Pierre Louis Guinand, a Swiss glassmaker, in 1805. He went to Munich, where he passed on his method to an apprentice optician, Joseph von Fraunhofer (1787–1826), who went on to make excellent refracting telescopes. Fraunhofer invented the equatorial mount, which allowed the telescope to move in any direction to focus on any part of the sky. It could be driven by a clockwork mechanism to move at a speed exactly matching the apparent motion of the stars, thereby tracking specific stars over the course of the night.

Although he died aged 39, Fraunhofer was the first to examine the spectral lines produced by stars. This was a major breakthrough and set the scene for the next chapter in the development of astronomy.

Lines in the dark and the light

The optical glass Fraunhofer made was among the finest in the world and was used in the best telescopes. He also made glass prisms. In 1814, Fraunhofer discovered that if he used one of his prisms to split light from the Sun he saw something more than just the expected spectrum of colours. The spectrum was crossed by thin black lines – by 547 thin black lines, in fact. He found that light from the stars also produced a spectrum crossed by black lines, but not in the same places as those in sunlight. He used his lenses to look at the flames from burning gases and discovered that some gases produced a spectrum with lines in the same places as some of those in the Sun's spectrum. But Fraunhofer did not investigate his discovery further – he was too busy making glass to take time to pursue a project in a science he knew little about – so for 50 years the spectral lines remained unexplained.

Then, in the 1850s, two German chemists set out to explore further the phenomenon that Fraunhofer had recorded. They were Robert Bunsen (1811–99), inventor of the Bunsen burner, and physicist Gustav Kirchhoff (1824–87), working together at the University of Heidelberg. They built a spectroscope, which consisted of a central prism and a set of miniature telescopes. Using a Bunsen burner to heat substances, they used the spectroscope to examine the light from the gases produced. They found that each produced a series of bright-coloured bands of light. These light bands, now known as emission spectra, were the inverse of the dark lines, or absorption spectra, that Fraunhofer had observed. By shining a background light through the flame, Bunsen and Kirchhoff could convert them to absorption spectra – a spectrum with dark lines that exactly matched the bright bands of the emission spectrum.

By investigating each element in turn in this way, the pair built up a reference bank of spectra – effectively, a spectral fingerprint for every known element. It was finally clear that Fraunhofer had discovered a way of investigating the chemical composition of the stars.

Lasting images

A spectroscopic catalogue requires photography, but this is not the only benefit of photography for astronomers. Photography revolutionized astronomy by offering the opportunity to record transient events accurately, preserving data to compare over time. The first attempt at photographing an astronomical object was

> 'At the moment I am occupied by an investigation with Kirchoff which does not allow us to sleep. Kirchoff has made a totally unexpected discovery, inasmuch as he has found out the cause for the dark lines in the solar spectrum and can produce these lines artificially intensified both in the solar spectrum and in the continuous spectrum of a flame, their position being identical with that of Fraunhofer's lines. Hence the path is opened for the determination of the chemical composition of the Sun and the fixed stars.'
>
> Robert Bunsen, 1859

EMISSION AND ABSORPTION SPECTRA

Emission lines appear when an electron in an atom drops into a lower orbit (closer to the nucleus) and loses energy. Absorption lines appear when an electron moves to a higher orbit and absorbs energy. The spacing and location of lines for different elements are distinct, so it's possible to identify an element from its spectra. By looking at the patterns in the wavelengths of electromagnetic radiation from celestial bodies astronomers can work out details of their composition, density, rotation and temperature.

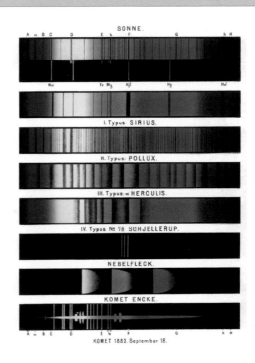

made by Louis Daguerre, who invented the daguerrotype process in 1839. His photograph of the Moon was a blurry smear, as keeping the equipment focused during the long exposure was too difficult. John Draper, a professor of chemistry in New York, had more luck the following year with a 20-minute exposure. The sun and a solar eclipse were successfully photographed in the following decade. The first photograph of a star was taken at Harvard Observatory in 1850, using an exposure of 100 seconds to produce an image of Vega, the second brightest star in the northern hemisphere. The first spectrogram of a star was photographed in 1863. With astrophotography in place, much else became possible.

Darkness visible

Light is only one part of the electromagnetic spectrum (see box on page 88) and represents only a small portion of the electromagnetic radiation emitted by stars. Because it is the portion that we can see, and that our ancestors have been able to observe for millennia, it has defined the way we think of the night sky and the universe. The disadvantage of modelling the universe around the objects that we can see – that emit visible light – is that we miss a lot of information, and even entire objects that emit other, non-visible, forms of radiation. Another disadvantage of relying on visible light is that we can only look at the stars at night-time. Working with different types of electromagnetic radiation, astronomers don't have to wait until after dark.

Light's place in a larger electromagnetic spectrum was only revealed by accident. In 1800, Herschel was investigating light, hoping to discover which colour of visible light produced most heat. Using a glass prism to split white light into its constituent colours, he recorded the temperature of each of the colour bands with a thermometer. He found that temperature increased from the blue end of the spectrum to the red end. When he placed the thermometer just past the red band, he recorded the highest temperature of all. He had discovered infrared.

The following year, German physicist Johann Ritter (1776–1810) discovered ultraviolet. He had set out looking for light (radiation) at the other end of the spectrum, and used photographic paper (paper coated with silver chloride) to do this. He found the paper turned black most readily when placed beyond the violet light in the spectrum.

The first detailed photograph of the full moon, taken by John William Draper in 1840.

The Scottish physicist James Clerk Maxwell (1831–79) made the first measurement of the speed of light and concluded there was little difference between light and electromagnetism. He wrote in 1864: 'light and magetism are affections of the same substance, and that light is an electromagnetic disturbance propagated through the field according to electromagnetic laws.'

This idea led him to predict the electromagnetic spectrum, which would be discovered in stages over the coming years. Heinrich Hertz discovered radio waves in 1886, providing the first proof of Maxwell's

---

ELECTROMAGNETIC RADIATION

Visible light is only one form of electromagnetic radiation – energy that is transmitted in waves and can travel through a vacuum. They are transmitted, simply, as vibrations in electric and magnetic fields. Light seems special to us because we can see it, but in reality it is just a small chunk of a continuum – or spectrum – that extends from radio waves at one end to gamma rays at the other. Radio waves have low frequency and a long wavelength; gamma rays have high frequency and a short wavelength.

Stars and celestial events such as supernovas (and the Big Bang, see page 184) generate electromagnetic radiation across the spectrum. This can be observed with radio telescopes and X-ray telescopes in just the same way that light can be observed with an optical telescope.

theory. Wilhelm Röntgen discovered X-rays in 1895 and Paul Villard observed gamma rays in 1900 while investigating radiation from radium. Neither X-rays nor gamma rays were immediately identified as further examples of electromagnetic radiation. X-rays found their place in 1912 and gamma rays in 1914, when Ernest Rutherford discovered they could be reflected in the same way as light.

Radio telescopes – the next era

It wasn't immediately obvious that other forms of electromagnetic radiation besides visible light could come from space. Indeed, it was an accidental discovery in 1931 that led to the age of radio astronomy. Radio engineer Karl Jansky (1905–50), working for Bell Telephone Laboratories in New Jersey, had built a radio antenna on a rotating platform 30m (98ft) wide. He was using it to investigate sources of potential atmospheric radio interference and categorized the types of static he detected as nearby and distant thunderstorms and a persistent hiss that he could not identify. It operated in the 20Hz band – a wavelength of about 15m (49ft). He came to the astonishing conclusion that the static hiss came from space. At

Karl Jansky at Bell Laboratories with his makeshift radio antenna, the first radio telescope, in 1933.

first, it seemed to come from the direction of the Sun, but after a few months it moved away. It was stronger and weaker on a daily cycle – but a sidereal day (relating to the Earth's position relative to the stars rather than the Sun). Eventually he concluded that the static hiss came from the direction of the centre of the Milky Way.

Bell Laboratories told Jansky to stop his investigations as they were not contributing materially to the company's plan to develop transatlantic voice transmissions. The baton was picked up by Grote Reber, a radio engineer in Illinois, who read Jansky's account of his findings which were published in 1933. Working independently, with no research funding, Reber built the first parabolic radio reflector telescope in his backyard. He decided to narrow the beam width from Jansky's 25-degree beam and chose a frequency of 3,000Hz, a wavelength of 10cm (4in).

The size of the reflector is important as it determines the resolution of the image. As visible light has a very small wavelength,

deciding the likelihood of life elsewhere in the universe. By mid-2016, 2,327 exoplanets had been confirmed.

What you see depends on how you look

If we could look at the night sky with eyes that saw a different band of the electromagnetic spectrum, such as X-rays or ultraviolet, we'd see a very different sky. Some objects which appear bright now would fade or disappear and objects which are invisible to our eyes would appear out of the darkness. The advent of observational equipment that can reveal the sources of different types of electromagnetic radiation beyond visible light has led to the discovery of new kinds of celestial objects and phenomena.

Going there

Until the 1960s, astronomy had to be conducted from within Earth's atmosphere. But with the advent of satellites and then space travel, our view of the stars and planets has changed. Exploration is no longer theoretical – an armchair tour of the cosmos with the help of telescopes. Now we can send robotic equipment, and even humans, to investigate what's out there. While we can learn a lot from observations at a distance, some details can only be gathered by a closer approach or a visit.

Bravely going where no man has gone before

Space travel and astronomy are not the same, just as the optical science involved in making telescopes is not the same as

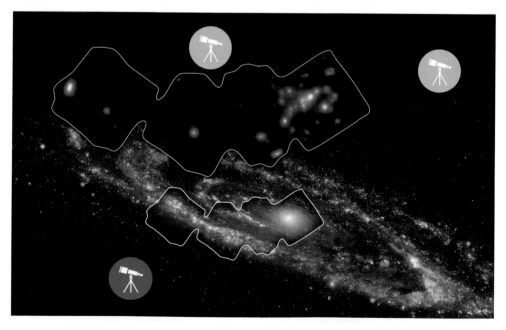

The Andromeda Galaxy, also known as M31, imaged by NASA using visible light (background image, blue icon), ultraviolet (pink icon) and X-ray (green icon) telescopes.

The cover of Jules Verne's De la terre à la lune *(From Earth to the Moon), 1865, one of the first modern science-fiction stories.*

astronomy. Space travel is, among other things, a tool of astronomy and cosmology. The presence of human astronauts or robotic probes enables us to find out more about the planets and other bodies and to piece together the history of the solar system and Earth.

The earliest probes took simple measurements, such as temperature and pressure, and pictures that provided spectroscopic and visual information. More sophisticated probes, such as those sent recently to Mars, can collect samples of gas, dust and rock and carry out chemical analysis and other investigations onsite. Data collected by probes is sent back to Earth by radio link; robotic probes are also controlled and reprogrammed by radio. There can be a delay of many minutes each way when communicating with probes on or near other planets.

AN OLD DREAM

Humans may have dreamed of visiting the stars for millennia. The first known 'science fiction' story about space travel, by the Syrian writer Lucian of Samosata, appeared in the 2nd century AD, though it was intended as a satirical dig at the fantastical travelogues then produced. His travellers are whisked to the Moon in a whirlwind and discover strange creatures and men (but no women) living there. In 1638, Kepler's fictional account of what it would be like to live on the Moon was published after his death. Both the science fiction writer Isaac Asimov and the astronomer Carl Sagan considered it to be the first work of science fiction.

The end of paradise

Venus was the first planet to be visited by probes. Early notions of what Venus might be like were beguiling. Being nearer to the Sun, it was thought likely to be pleasantly warm. The planet's cloudy atmosphere was taken to indicate hot, steamy conditions. It was, perhaps, a tropical paradise. It was often depicted this way in science fiction, and Soviet craft sent to Venus in the 1960s were equipped to land on water if necessary.

With the data returned from the planet, the dream came to an end. NASA's Mariner 2 probe found on a flyby mission in 1962 that the ground temperature on Venus could reach 428°C (802°F) and there was no evidence of water vapour in the atmosphere. This was confirmed by the Soviet Venera probes that landed on the planet. Instead of a tropical paradise, Venus turned out to be a scorching hot, dry desert with crushing atmospheric pressure and rather too many

acidic clouds for comfort. In 1967, Venera 4 returned the first data from another planet, reporting on temperature and pressure during its descent, but was destroyed by atmospheric pressure before reaching the surface. In 1970, Venera 7 sent data from the surface; it found a temperature of 475°C (887°F) and pressure of 90 atmospheres. The hostile nature of Venus was confirmed. Science fiction writers Brian Aldiss and Harry Harrison marked the passing of the Venusian fantasy with an anthology entitled *Farewell Fantastic Venus* (1968).

Probes have since photographed the other planets from nearby, and many of their moons. These close-up photos have revealed the likely composition and structure of some of the bodies studied, teaching us far more about them than could have been garnered from Earth. In 2015, NASA's Cassini mission to Saturn and its moons found that there is likely to be a globe-wide ocean of liquid water under the ice crust of the moon Enceladus. In 2104, the same mission found that Titan, another moon of Saturn, has vast seas of liquid methane. This could only be detected using radar from a nearby space probe. Cassini has also collected tiny grains of dust, including, in 2016, 36 specks that have come from outside the solar system. Again, it's something that could never have been detected from Earth.

Dust and rocks

The crowning achievement of the 1960s space programmes came on 20 July 1969 when Neil Armstrong and Buzz Aldrin, two of the three-man crew of Apollo 11, stepped out onto the Moon. Most important for

---

NEWTON LOOKS AHEAD

In 1687, Isaac Newton calculated that if a cannonball were fired from the top of a mountain at a speed of 7.3km (24,000ft) per second, it would go into orbit around the Earth, as it would be falling towards Earth as fast as Earth was moving away beneath it. If the cannonball were fired at an even greater velocity, it would break free from Earth's gravity and head off into space. This speed has become known as escape velocity. From Earth, escape velocity is 40,270kmh, or about 11km (7 miles) per second.

In 1975, the Soviet space probe Venera 9 transmitted the first-ever photograph from the surface of another planet, visible evidence that Venus was not the tropical paradise people had dreamed of.

astronomy was not the presence of the humans there but the 2,200 samples of moon rock and dust collected, totalling 382kg (842lb). Analysis of this helped to shed light on the formation of the Moon (see page 106).

Other samples have been returned from comets and collected on (but not returned from) Mars. Landings on Mars began with Soviet craft Mars 2 and Mars 3 in 1971. Since then, missions to Mars have become increasingly sophisticated. The first successful landing of a rover (robotic vehicle) on Mars was Sojourner in 1997. It had cameras and a spectrometer to analyse samples of rocks and dust. The most recent Mars rovers (from NASA) continue to send back data from geological and climate/atmosphere investigations, plus millions of photographs. In 2014, the rover Curiosity collected the first sample ever taken by drilling into rock not on Earth. The aim of NASA's current generation of Mars rovers is to determine whether there has ever been life on Mars, and the planet's potential for supporting life.

Plumes of water ice erupting from the surface of Enceladus, one of Saturn's moons, suggest that there is liquid water below the surface.

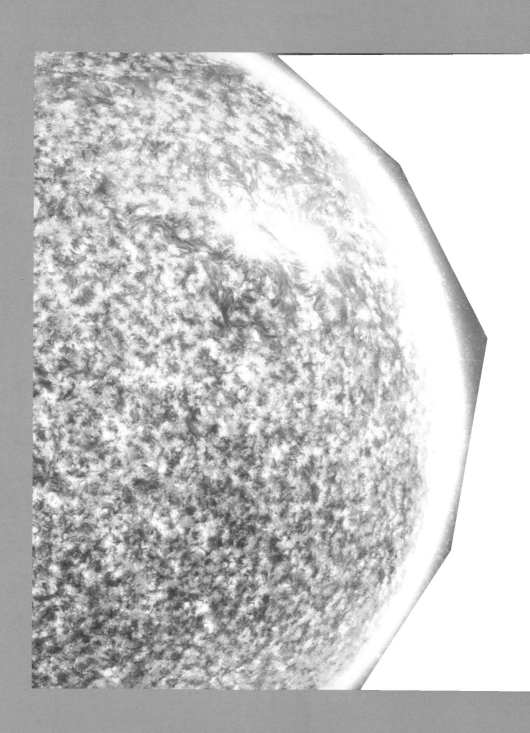

CHAPTER 4

Earth, Moon
AND SUN

'The sun comes into being each day from little pieces of fire that are collected.'

Xenophanes,
Greek philosopher,
c.400BC

It's easy to think that astronomy is about objects 'out there' in space, planets, stars, comets and so on, and to forget that our own planet is part of it all. The closest bit of 'out there' – our own little system of the Earth, Moon and Sun – was the first to be explored.

The Sun and Moon are the bodies that have most impact on the Earth. They appear largest in the sky, and first attracted our ancestors' curiosity.

The Earth in space

While determining whether the Earth or Sun was at the centre of the local system (or even the entire universe) was very difficult, some tasks were more easily accomplished. The size of the Earth, its sphericity, and its distance from the Sun and Moon were all challenges that could be met using relatively simple measurements and mathematics.

The round Earth's imagined corners

There's a popular belief that many people thought the Earth was flat for a very long time, and that the sea captains of Columbus's era were afraid of dropping off the edge of the world. But this is unfounded; there is plenty of evidence for the sphericity of the Earth.

It would have been obvious to any early mariner that the Earth is not flat. Ships at sea do not recede until they are mere dots, but disappear over the horizon while still close enough for details such as their sails to be clearly visible. The fact that the hull of a ship disappears first, followed by its sails, is proof that the Earth's surface is at least curved, although early cosmological models imagined the Earth to be domed rather than spherical.

It's easy to see that the Earth has a curved surface, but it's not immediately apparent that it's a sphere. Two thousand years ago, the known land comprised a continuous landmass and a few scattered nearby islands. Travelling from Greece or the Middle East, you would have had to go a very long way south or east before running out of land and meeting the ocean. You

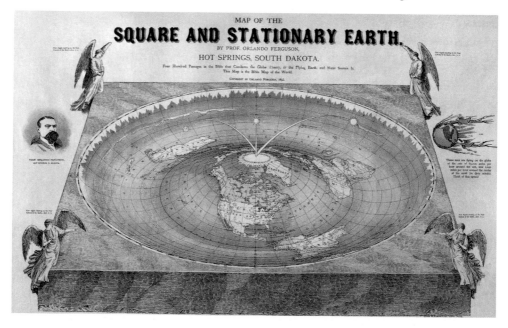

This flat Earth map was drawn by Orlando Ferguson in 1863; he claimed it represented the true Christian 'map of the world' and supported it with quotations from the Bible.

A lunar eclipse can happen only when there is a full moon, as the Sun, Earth and Moon are aligned. The shadow of the Earth falls over the Moon.

would have encountered impassable desert or mountains first. Going north or west would have led to the Atlantic Ocean or frozen north fairly quickly, suggesting, as all early maps proposed, that the inhabited land is surrounded by inhospitable land (very cold or very hot) and then an ocean that marks the edge of the world. (The Mediterranean Sea was easily skirted around or navigated by the time of the Ancient Greeks.)

The Greeks are the first people known to consider the Earth to be spherical, though it was not unanimously held to be true. Aristotle argued in the 4th century BC that the Earth must be spherical because the shadow it casts on the moon during a lunar eclipse is always circular. (A lunar eclipse occurs when the Earth lies between the Sun and the Moon, so that the shadow of the Earth falls across the Moon.)

Further, Aristotle argued that the Earth is not of enormous size because it is easy to travel far enough to the north or south of Greece to see differences in the stars. Some that are visible in the north throughout the year seem to rise and fall further south, and some not visible at all in the north are seen in the south. The idea of a spherical Earth did not originate with Aristotle, but he is the first to leave an argument for it. Other Greek writers attributed the original discovery to Pythagoras, Parmenides (late 5th/early 6th century BC) or Empedocles. One Greek philosopher, Archelaus (5th century BC), proposed that the planet is saucer-shaped, with a dip in the centre, as otherwise (if it were flat) the Sun would rise and set at the same time everywhere. Archelaus also suggested that the Sun is the largest of the stars.

How far round?

The Greek philosopher Eratosthenes was first to demonstrate the sphericity of the Earth conclusively in the 3rd century BC and to make the first attempt at calculating its size. Eratosthenes lived in Alexandria in Egypt and was librarian of the great Library of Alexandria. He heard that in Syene, a town to the south, the Sun was directly overhead at noon. It shone straight into a deep well, illuminating the bottom, and vertical objects cast no shadow. Eratosthenes knew that in Alexandria this was not the case: vertical objects cast a small shadow at noon. He realized that if he could measure the distance between the two cities and the angle of the shadow in Alexandria, he could work out the angle of the Sun at Alexandria when it was overhead at Syene and so calculate the curvature of the Earth.

Eratosthenes found the angle to be one fiftieth of a full circle (about 7.2 degrees), so the distance between Syene and Alexandria represented one fiftieth of the circumference of the world. He reported the distance between the cities as 5,000 stadia, making the circumference 250,000 stadia. He then pushed the measurement up to 252,000 stadia to make the figure easier to divide by 60 (he knew his value was approximate). Unfortunately, we don't know the exact length of a stadium. If he meant an Egyptian stadium, his calculated value of 250,000 stadia is only a few percent off the recognized circumference of 40,075km (24,860 miles). In any case, his error falls between 2 and 20 percent, which is impressive.

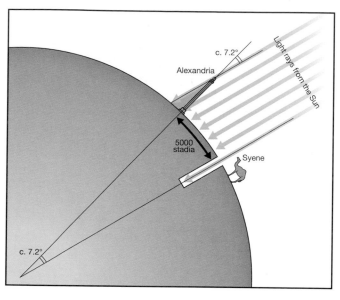

Erathosthenes' method of measuring the circumference of the Earth: the angle subtended at the Earth's centre between radii drawn to Alexandria and Syene is the same as the angle formed by the Sun falling on the pillar at Alexandria.

In the 9th century, frustrated by their inability to convert the values in unknown stadia to Arab miles, Islamic astronomers set out to recalculate the size of the Earth. A team went into the desert of Syria and chose a bright star for their observations. Then half the team set out going due north and the other half due south. Each team stopped when they found that the position of the chosen star had moved by one degree. The distance between the two points gave the circumference corresponding to two degrees of arc, so the total circumference was 180 times that distance. The resulting calculation gave the circumference of the Earth as 38,624km (24,000 miles), which is close to the modern measurement.

Going round

As we have seen, opinion was divided in Ancient Greece about whether the Earth rotates on its axis or whether the celestial bodies move around it. The Earth's own rotation is an issue separate from whether or not our own planet revolves around the Sun. Several Greek philosophers of the 4th century BC, including Hicetas, Heraclides and Ecphantus, maintained that the Earth rotates, but they did not suggest that it orbits the Sun. Aristotle was adamant that the Earth was fixed and as his view was adopted and propagated by Ptolemy it became the dominant belief in the West until the Copernican model replaced it.

In India, Aryabhata wrote in AD499 that the Earth rotates daily on its axis and the apparent movement of the heavenly bodies is the result of that rotation. It was a question discussed by Muslim astronomers, too, though no consensus was reached.

With Copernicus, the question of rotation became more compelling. Copernicus required the Earth to turn as, otherwise, in his model there would be extended periods of darkness, light and twilight in a cycle that took an entire year to complete. Some people accepted the rotation of the Earth without committing to either a geocentric or heliocentric model.

In 1687, Isaac Newton established that if the Earth rotates, the poles will be slightly flattened and the Earth will bulge at the equator. Early measurements late in the same century suggested this was not the case – although in fact it is. The French Geodetic Mission in the 1730s vindicated Newton and the Copernican model, finding that the Earth is indeed an oblate spheroid, slightly squashed.

The ultimate proof that the Earth turns was provided in 1851 by French physicist Léon Foucault (1819–68). He hung a free-

Hanging 67m (220ft) from the roof of the Panthéon in Paris, France, Foucault's Pendulum makes it possible to see the Earth turning.

swinging ball of lead, coated with brass, from the roof of the Panthéon in Paris; it is still there. Because the Earth rotates beneath the pendulum, the plane of its swing slowly rotates. Watching it for a few minutes is enough to see that the Earth moves; the plane of swing moves 11 degrees per hour, or about one degree every five minutes.

Our companion, the Moon

The Moon and Sun are both clearly round – everyone can see that – but establishing that they are spherical rather than circular discs is a little harder. The Greek philosopher Heraclitus suggested around 500BC that the Moon and Sun are bowls of fire. The Moon's bowl rotates, he said, so from Earth it appears to be different shapes at different times, accounting for its phases.

Eclipses result from the convex side of the bowls facing the Earth. He also thought evaporation from the land and sea provided fuel for the heavenly bodies, which burned in the same manner as oil lamps.

Reflected glory

The Greek philosopher Anaxagoras (c.510–428BC) believed both the Moon and the Sun to be round lumps of rock; he also held that the Moon is not a source of light itself, but shines with reflected sunlight. This is fairly easy to deduce from the phases of the Moon (see diagram). When the Moon is between Earth and the Sun, the side we see is in shadow (the new moon); when Earth is between the Sun and the Moon, we see the Moon fully illuminated by the Sun (the full moon); between these, the

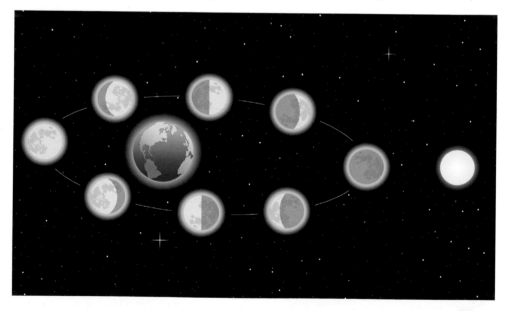

Sunlight falls onto the side of the Moon that is facing the Sun. How much of that part can be seen from Earth depends on the position of the Moon in its orbit around the Earth. At full moon, we can see the whole of the illuminated side; at new moon, we can see only a thin sliver of light.

other phases of the Moon show it partially illuminated.

Chinese astronomers also recognized that the Moon shines with reflected light and that both Moon and Sun are spherical. Jing Fang (78–37BC) promoted the theory, though it was not universally accepted. In India, too, Aryabhata noted in AD499 that the Moon shines with reflected light.

The face of the Moon

It's clear even to a naked-eye observer that the Moon's surface is not uniformly lit; there are dark (shadow) areas which have often been imagined to show a human face or other figure. The Greek historian Plutarch wrote a short treatise called *On the Face in the Moon's Orb* in which he suggested that the dark areas on the Moon's surface are really pits and chasms, perhaps rivers, which are too deep for the Sun's light to fall into. He was even willing to consider that the Moon might be inhabited.

The earliest surviving drawing of the Moon's features was made by Englishman William Gilbert (1540–1603), physician to Queen Elizabeth I. In an inversion of the model adopted by Plutarch and, later, Galileo, Gilbert assumed that the dark spots were continental landmasses and the light areas seas. Gilbert reflected that it was disappointing that no one in antiquity had drawn the face of the Moon, as this oversight meant it was impossible to tell whether it had changed in the last 2,000 years. Gilbert did not publish his image of the Moon, and it did not appear in print until 1651, long after his death. By that time, of course, it had been entirely superseded by maps of the

THE KNOWTH MOON MAP
The oldest possible depiction of the Moon's surface is on the wall of a cave in Knowth, Ireland and might date from 5,000 years ago. The stone carving, known as Orthostat 47, is in a Neolithic passageway and its pattern of curved lines seems to correspond to the arrangement of the dark patches (now known as maria, or seas) on the Moon.

Moon made with the aid of a telescope. It remains the only historic map of the Moon drawn from naked-eye observation.

With the invention of the telescope in 1608, the Moon was compelled to give up its secrets. The first person known to turn a 'Dutch trunke' (an early name for a telescope) towards the Moon and draw what he saw through it was the English astronomer Thomas Harriot (1560–1621). He drew his first picture of the magnified Moon in July 1609, several months before Galileo made his own first drawings of the Moon. Harriot's simple sketch shows some crude features of the surface on the lighted portion of the

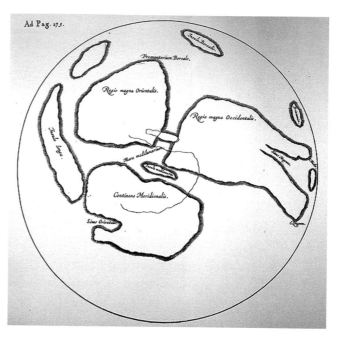

Gilbert's drawing is the only surviving map of the Moon made before the invention of the telescope.

Moon, and the terminator (the line that divides the part of the Moon lit by the Sun and the part in shadow).

When Galileo made his telescope in 1609, and turned it towards the Moon, he realized that it was not, as previously supposed, a perfect, smooth sphere but 'on the contrary, it is full of great

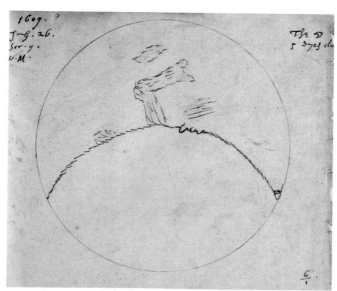

Harriot's drawing of the Moon, the first made with the use of a telescope, clearly shows the terminator dividing light and dark parts.

inequalities, uneven, full of hollows and protuberances, just like the surface of the Earth'. This was a blow, as both Aristotle and the Church taught that the heavens are perfect and unchanging. The Moon, clearly, was not so perfect, even if it was unchanging. Of course, darker patches have always been visible on the Moon, but these were deemed to be the result of unequal lighting from the Sun rather than surface imperfections. In 1824, the German astronomer Franz von Gruithuisen correctly deduced that the Moon's craters were the result of meteors crashing into the surface – so the Moon was not even unchanging.

Mapping the Moon

The first proper map of the Moon was made by Belgian astronomer and cosmographer Michael van Langren in 1645. It was superseded in 1647 by the more influential maps of Johannes Hevelius, who built a huge wooden telescope with a focal length of 46m (151ft) which he installed in an observatory across the roofs of three adjacent houses he owned. Even so, Hevelius preferred

to work without a telescope, and is considered the last major astronomer to have carried out significant naked-eye observations. He did use one, though, for his four-year project to map the lunar surface. He published *Selenographia*, his full description of the Moon, in 1647. Just four years later, the Italian astronomers Giovanni Riccioli and Francesco Grimaldi drew a new map of the Moon, giving names to many of the craters. Their names are still in use today.

Up to this point, the maps of the Moon were freehand drawings with no reference system. When Johann Meyer developed the first set of lunar coordinates in 1750, astronomers were able to map features on the Moon systematically, a process which began officially in 1779 with the

Michel van Langren made the first map of the Moon in 1645, produced to help mariners find longitude at sea.

work of the German amateur astronomer Johann Schröter (1745–1816). Fortunately, Schröter published his detailed drawings of the Moon's surface before his papers and observatory were destroyed by the French during the Napoleonic wars. He also drew images of Mars, but believed that he was observing clouds in the Martian atmosphere rather than features of the planet's surface.

Mapping the Moon is now carried out using sophisticated technology and highly advanced telescopes. The most accurate and detailed topographical map of the Moon, released in 2011, was produced by NASA and based on photographs.

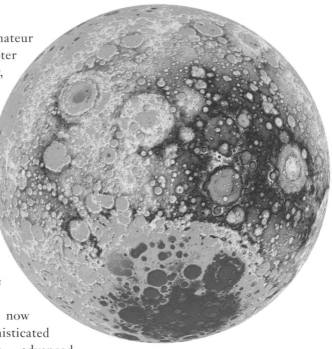

NASA's false-colour image is the highest resolution map of the Moon ever created. Highest elevation is white, then red, going through the spectrum to purple at the lowest elevation.

THE QUESTION OF LIFE

The idea that there might be life on the Moon has been debated for a long time and was still current among some astronomers as late as the early 20th century. Originally the preserve of fiction writers and purely speculative astronomers, the notion was given a boost in 1856 when the Danish-German astronomer Peter Hansen (1795–1874) published the theory that the Moon might have an atmosphere and be able to support life on its far side (which is always facing away from Earth). The Croatian polymath Roger Boscovich (1711–87) had discovered in 1753 that the Moon has no atmosphere and two German astronomers had shown in 1834 that it has neither atmosphere nor water. Hansen's theory was discredited in 1870, but not all astronomers were keen to give up the idea of life on the Moon. Speculation about past or present life continued until the Apollo moon landings in the 1960s returned sterile moon rock and confirmed the absence of liquid water on the Moon. But then, in 2009, NASA reported that there is water on the Moon after all – enough to supply human needs on any future missions.

All the way to the Moon

That the Moon is closer to the Earth than the Sun is evident from the occurrence of eclipses. In the 5th century BC, Anaxagoras correctly explained that an eclipse is the result of the Moon passing in front of the Sun. Beyond that simple sequence, it was impossible to judge just how far away the Moon (or Sun) might be. Both look about the same size in the sky, and the Moon can perfectly cover the Sun during an eclipse. This could mean they are of similar sizes and distances, with the Sun just slightly further away, or (as the Sun is much brighter than the Moon) it could suggest that the Sun is further away and larger. The first person to apply himself to the calculation was the Greek astronomer Hipparchus in the 2nd century BC. He used the method of parallax (see box). All indirect methods of measuring the distance between the Earth and Moon use parallax in one form or another. It requires a simple measurement on Earth between two points, together with some geometrical calculations. Parallax works by observing the same object from two different positions and determining how much it appears to shift against a constant background. The closer the object, the more it seems to shift.

PARALLAX

Parallax is the apparent change in position of an object, including a star or planet, that results from changing the viewpoint of the observer. You can demonstrate parallax by holding up a finger and closing first one eye and then the other. The finger appears to jump to the right or left against the background as you observe it from two slightly different places – the position of your two eyes. In astronomy, parallax is measured as the angle of incline of the line of sight; the closer the object, the larger the parallax angle. For celestial bodies, which are far away, this angle is very small. Although it is possible approximately to measure the parallax of the Moon without a telescope, the parallax of the stars was too small to be measured until the 19th century (see page 187).

Hipparchus made his measurements during an eclipse, quite possibly in 129BC. The eclipse was total at the Hellespont (now known as the Dardanelles), which divides the European and Asian parts of Turkey, but only four fifths of the Sun was covered at Alexandria in Egypt. Hipparchus

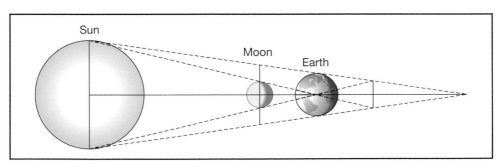

Hipparchus used geometry to determine the distance from Earth to the Moon and Sun.

probably did not work from the distance between the two cities, but the difference in their latitude (now known to be about 9 degrees). As the Sun usually occupies about 0.5 degrees of the whole arc of the sky, a fifth of the Sun occupies 0.1 degrees. From this, Hipparchus worked out from the change in latitude that the distance from the Earth to the Moon was between 70 and 82 times the radius of the Earth. The distance is about 60 Earth radii, so Hipparchus' calculation was a reasonable first attempt.

The Persian astronomer Habash al-Hasib al-Mawarzi (796–c.869) made a more accurate calculation of the distance between the Earth and the Moon. He gave the distance to the Moon as 346,345km (215,209 miles) – the actual figure is 384,400km (238,855 miles) – and calculated the Moon's diameter as 3,037km (1,887 miles) – the actual figure is 3,474km (2,158 miles).

Another way of judging the distance between the Earth and the Moon involves measuring meridional transits. This requires observers widely separated on the same line of longitude (on a meridian) to observe the passage (transit) of a feature of the Moon over that line. The Irish astronomer Andrew de la Cherois Crommelin (1865–1939) calculated the distance to the Moon from measurements of the angle of the Moon's elevation made in 1905–10. The line of transit ran between Greenwich in England and the Cape of Good Hope in South Africa. Crommelin reported a distance to the Moon that was accurate to ± 30 km (18½ miles); it remained the accepted value for 50 years.

In 1952, John O'Keefe (1916–2000) measured the angle of elevation of the Moon from different places as the Moon occluded (hid) a particular star. Their value for the distance was 384,407.6 ±4.7km (238,859.8 miles, +/- 2.9 miles), refined in 1962 by Irene Fischer to 384,403.7 ±2km (238,857.4 miles, +/- 1.2 miles).

Today we measure the distance from Earth to the Moon directly, using lasers. A laser beam is bounced off an object on the Moon, and the time it takes for the reflected beam to return to Earth is recorded. The technique was first used in 1962, bouncing a laser beam off the Moon's surface. In 1969, the crew of Apollo 11 placed a retroreflector array on the Moon specifically for this purpose. This device reflects laser light back to its source with the minimum of deviation or distortion, and thus gives the most precise measurement possible.

How big?

In the 3rd century BC, Aristarchus used a lunar eclipse to calculate the size of the Moon. He recorded how long it took from the start of the eclipse (when the Earth's shadow first fell on the Moon) until the Moon was fully hidden, and for how long the Moon was totally hidden. Finding the two times to be the same, he concluded that the Earth's shadow must be twice the diameter of the Moon, so the Moon must be half the size of the Earth. In fact, the Moon is about a quarter the size of the Earth. Aristarchus had assumed that Earth's shadow was the same size as the Earth, whereas in fact the shadow is a good deal smaller.

What Aristarchus thought the shadow was like:

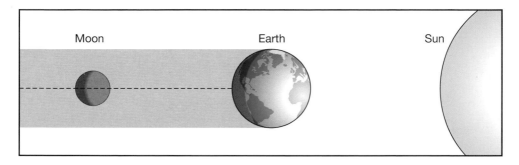

What the shadow is actually like:

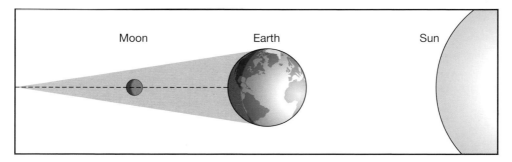

Creating the Moon

It's one thing to observe and investigate the Moon, but quite another to wonder why it is there at all. The first person to attempt to give a scientific explanation for the presence of the Moon was the English astronomer and mathematician George Darwin (1845–1912), second son of the famous English naturalist Charles Darwin. In 1898 Darwin proposed that the Earth and Moon had once been a single body, but the Moon had spun off the Earth as a globule of hot, molten rock cast away by centrifugal force. He said that the Moon was still moving away from the Earth, which was confirmed by laser ranging measurement of the Moon in the second half of the 20th century. (The Moon is moving away from Earth at a rate of about 3.8cm (1½in) per year.)

Mathematical modelling did not support Darwin's theory, though, and in 1946 the Canadian geologist Reginald Daly (1871–1957) suggested that, instead, the Moon had been blasted out of the Earth by an impact. The idea was largely ignored until 1974. Then William Hartmann and Donald Davies suggested that, soon after the planets were formed, a few large bodies existed which might either be captured by planets as satellites or collide with them to devastating effect. Their idea, now known as the Giant Impact Theory, was that one such body had crashed into Earth, throwing a large amount of material into space which had coalesced into the Moon. The hypothetical colliding body, possibly the size of Mars, was designated Theia in 2000. The collision is thought to have happened 100 million years after the solar system began to form, making the Moon 4.53 billion years old (to the Earth's 4.54 billion years). Analysis of rocks and dust returned from the Moon by the Apollo missions in the 1960s and 1970s supports the theory. The samples are of similar composition to the Earth, but not identical with it. It seems that the Moon combined parts of Theia and parts of Earth.

The Sun

Even though the Sun is seen in daytime and the stars at night, the idea that the Sun might be a star first arose nearly 2,500 years ago. This was not generally accepted, though, until proposed by Galileo in the 17th century.

TIME AND TIDE . . .

The first person to realize that the tides are caused by the Moon was Seleucus of Seleucia, a Hellenistic astronomer living in Mesopotamia in the 2nd century BC. He studied the tides, noting that they were not entirely regular but varied in height depending on the Moon's position in relation to the Sun. He thought them mediated by the 'pneuma' (breath), of which the universe was originally composed (see page 182), and produced by both the Moon and a whirling motion of the Earth.

SCARING THE DRAGON

Ancient Chinese mythology held that an eclipse was caused by a dragon eating the Sun. It became common to make loud noises, such as beating drums, to scare it away. Up to the 19th century, the Chinese navy fired cannons during a lunar eclipse to scare away the dragon that was eating the Moon.

Now you see it, now you don't

The dependability of the Sun was no doubt a source of comfort to our ancestors. As we have seen, they arranged their calendars around it and built structures that enabled them to use its rising and setting to determine suitable times for activities such as planting and harvesting. How terrifying it must then have been when the Sun behaved in unexpected ways – when solar flares varied its appearance or, worse yet, when it was obliterated or part eaten away by an eclipse. Fear was a natural response, and superstition followed close on its heels.

In both Ancient China and Babylon, eclipses were believed to be a bad omen, particularly for rulers. Activity in the heavens was thought to relate to national and political events on Earth rather than to the minutiae of everyday lives. In this regard, modern astrology has moved far from its roots. In Babylon, a substitute ruler might be recruited to perform duties to pacify the gods and avert the coming misfortune. In one case, a prediction that a flood would break through the dykes led to this proffered remedy: 'When the Moon has made the eclipse, the king, my lord, should

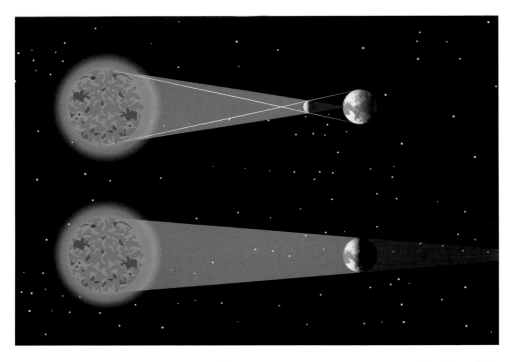

In a solar eclipse (top) the shadow of the Moon falls on the Earth. In a lunar eclipse (bottom) the shadow of the Earth falls on the Moon (just visible on the dark side of the Earth).

write to me. As a substitute for the king, I will cut through a dyke, here in Babylonia, in the middle of the night.'

Recording eclipses

A stone in Ireland may record an eclipse that took place on 30 November 3340BC. If

> 'These late eclipses in the sun and moon portend no good to us. Though the wisdom of nature can reason it thus and thus, yet nature finds itself scourged by the sequent effects.'
>
> William Shakespeare,
> *King Lear*, Act I, Scene 2

that interpretation of the stone is correct, it would be the earliest eclipse recorded. A Syrian clay tablet records an eclipse on 5 March 1223BC, which would otherwise be the earliest western record.

Chinese records of eclipses go back 4,000 years and are sufficiently complete to have been used to calculate changes in the Earth's rate of spin. The earliest surviving account of an astronomical observation relates to an eclipse that occurred on 5 June 1302BC:

> 'Fifty-second day: fog until next dawn.
> Three flames ate the sun and big stars were seen.'

The 'three flames' are streamers within the atmosphere of the Sun that are seen stretching out into space during the eclipse. The 'big stars' are the bright stars and planets seen (in their unfamiliar daytime positions) during the darkness of the eclipse. The eclipse is preserved on a Chinese oracle bone (a piece of turtle shell). Five solar eclipses recorded by Chinese astronomers between 1161BC and 1226BC have enabled scientists from NASA to determine that the year was about 17 seconds shorter in 1200BC than it is now. Shi Shen gave instructions for predicting eclipses that were based on the positions of Sun and Moon, showing that in the 4th century BC he knew something of the part the Moon plays in a solar eclipse. Chinese astronomers understood the full nature of eclipses by 20BC and could predict them accurately by AD206.

> 'It was a total eclipse of about 12 digits or points. Also, such darkness arose over the Earth at the time of mid-eclipse that many stars appeared. No doubt this portended the very great and destructive calamities which were soon to be vented on the Romans by the Turks.'
>
> Eclipse seen at Constantinople (Istanbul), 25 May 1267, in *Nicephori Gregorae Byzantinae Historiae*

Predicting eclipses

From earliest times, predicting eclipses became one of the tasks assigned to astronomers in Mesopotamia and China. It seems that both cultures noticed the existence of the so-called Saros cycle. The geometry of the movements of the Sun, Earth and Moon repeats on a cycle of 6585.32 days (18 years, 11.3 days). Each Saros series continues for 1226–1550 years, with the zone of totality of the eclipse gradually moving from one pole to the other. This means that once an eclipse has been noted, it can be predicted with confidence that another will happen 6585.32 days later. The extra third of a day was a nuisance, as it meant that the eclipse would not be in the same place, but a third of the way around the world from the position of the first. Even so, it would return to the same place on every third cycle. There are 42 Saros cycles running at any one time. This means there are actually quite a lot of eclipses, just not very many that are visible from one place. The Saros cycles were first recorded on clay tablets in Babylon. Astronomers didn't need to understand what was happening to be able to predict eclipses from the cycle; they only needed sufficient data to spot the pattern and extrapolate from it.

As we have seen, the Greek philosopher Anaxagoras correctly explained eclipses in the 5th century BC – but that did not defuse their power to alarm. It's not entirely certain which eclipse Thales predicted to give the Medes an advantage in a battle against the Lydians (see page 35), but it is commonly said to have been that on 28 May 585BC. There is no record of how Thales was able to predict the eclipse, but as he was familiar with Babylonian astronomy it was possible that he made use of a Saros cycle.

Accurate prediction of eclipses came much later; the English astronomer Edmond Halley (of Halley's comet) predicted the

A map drawn by Edmond Halley showing the path of the shadow of the Moon over England during the eclipse of 3 May 1715. The total eclipse was visible over an area stretching from Kent to York in the east and from Cornwall to Wales in the west.

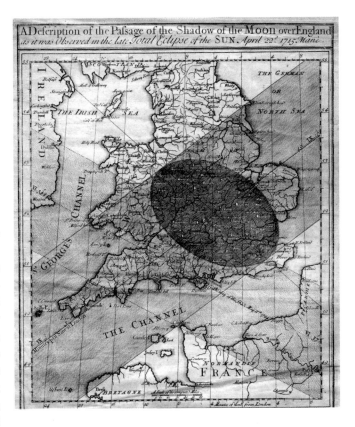

A Description of the Passage of the Shadow of the Moon over England as it was Observed in the late Total Eclipse of the SUN April 22d 1715. Mane.

eclipse of 3 May 1715 to within four minutes, and its path accurately to within 30km (19 miles).

Even when eclipses could be predicted and explained, they continued to be a source of fear and were associated with misfortune. Unfortunate coincidences, such as the death of King Henry I just after a total eclipse seen on 2 August 1133, tended to reinforce this connection. However, when Mohammed's followers claimed that an eclipse that coincided with the death of the prophet's son Ibrahim was a miracle, Mohammed denied it, saying that eclipses bear no relation to the dealings of men at all.

> '[Thales] says that eclipses of the sun take place when the moon passes across it in a direct line, since the moon is earthy in character; and it seems to the eye to be laid on the disc of the sun.'
>
> Aëtius of Antioch,
> 1st–2nd century BC

Sizes and distances

The Greek mathematician Aristarchus was the first to try to calculate the distance from Earth to the Sun. He realized that when the half-moon was visible, the trio of Earth-Sun-Moon formed a right-angled triangle. In his only surviving work, *On the Sizes and Distances*, written in the 3rd century BC, he estimated the angle between the Moon and Sun and calculated the ratio of the distances between the Earth and the Sun and between the Earth and the Moon. Judging the angle to be 87 degrees, he concluded that the Sun is 19 times as far away as the Moon. Unfortunately, his estimate of the angle was slightly out – the real angle is 89 degrees,

USEFUL ECLIPSES

Eclipses provide opportunities to make observations or measurements that would not otherwise be possible. One of the first people to do so was probably Hipparchus, who calculated the distance to the Moon, based on the Moon's parallax measured during an eclipse in 129BC (see page 107).

The element helium was discovered by the French astronomer Jules Janssen (1824–1907) observing the spectrum of the Sun during a total eclipse on 18 August 1868. Helium is the second most abundant chemical element in the universe (24 per cent), but is very rare on Earth. It was the first element to be discovered in space, before being found on Earth.

In 1919, observations by the English astronomer Arthur Eddington (1882–1944) of a total eclipse on the island of Principe, near Africa, confirmed part of Albert Einstein's theory of general relativity (see page 179). Eddington observed light from stars too close to the Sun to be seen normally, but visible during an eclipse. He was able to prove that the Sun's gravitational field bends light from the stars so that they appear to be in a slightly different position from their actual one (see page 181). This effect is called gravitational lensing and is now used extensively by astronomers.

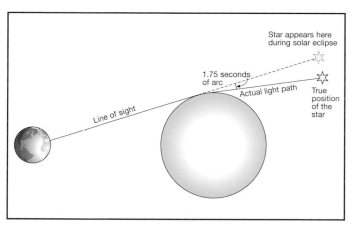

Star appears here during solar eclipse

1.75 seconds of arc

Actual light path

True position of the star

Line of sight

Since the Sun and Moon appear to be the same size in the sky, yet the Sun (he believed) was 19 times further from Earth than the Moon, Aristarchus concluded that the Sun must be 19 times larger than the Moon.

51 minutes. The small error scales up to a massive difference in distance: the Sun is 400 times as far away as the Moon. Even so, his conclusion that the Sun is much further away from Earth than the Moon was significant and valid. His estimate of the distance was accepted for around 2,000 years, until good-quality telescopes made more accurate measurements possible by parallax.

Hot rock or burning gas?

The idea that the Sun is fiery is intuitive. It shines brightly, it radiates heat – most

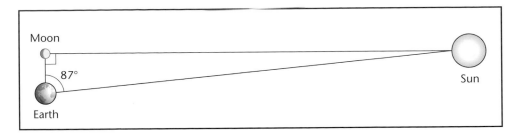

Aristarchus estimated the angle at Earth between the Sun and Moon, which formed the basis of his calculation of the distance to the Sun from the Earth.

efficiently when not obscured by clouds – and can even start fires on Earth. But a fire needs fuel and it is not at all clear, in the case of the Sun, what this might be.

Around 450BC, Anaxagoras proposed that the Sun is actually a star. This represented a considerable imaginative leap, requiring a sophisticated perception of three-dimensional space in which the other stars are very far away. He thought that the stars are fiery stones, and the Sun is a star so close that it appears much larger than the others so we can feel its heat on Earth. He even made an attempt to calculate the size of the Sun, putting it at larger than the Peleponnese, a large peninsula in Greece (so greater than a few hundred kilometres across).

Establishing a pattern that would become all too common, Anaxagoras was imprisoned and charged with promoting views that contradicted prevailing beliefs – namely, that the Sun is not a god, but a hot rock. Sentenced to death or exile by the Athenian court, he left the city and lived the rest of his life in Lampsacus in the eastern Hellespont.

Aristarchus also suggested that the Sun is simply a star up close (or the stars are suns at a distance), but, again, the idea didn't catch on.

Eclipse tables in the Dresden Codex, which show Mayan methods for calculating or tracking eclipses. The eclipse data is on the three lefthand panels.

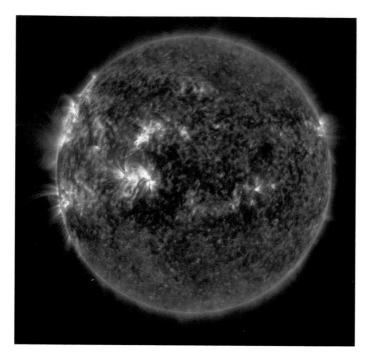

NASA's images of the Sun taken from the Solar Dynamics Observatory reveal a surface in constant flux and turmoil.

Powering the Sun

Until quite recently, the Earth and Sun were thought to be relatively young. In 1654, the Irish Protestant Archbishop James Usher (1581–1656) calculated the age of the Earth by adding together the ages of all the patriarchs in the Old Testament. He came up with a date for Creation of 23 October 4004BC, making the Earth (and therefore the Sun) less than 6,000 years old at the time of his calculation. But in the 18th and 19th centuries, advances in geology and evolutionary theory pushed the starting point for the Earth much further back – first to around 100,000 years and then to millions of years.

This led to renewed interest in the source of the Sun's energy and how long it might last. It was clear the Sun was not simply burning fuel in a way familiar on Earth. No normal chemical reaction could be sustained long enough to power the Sun for millions of years.

One theory was that the Sun's gravity attracted meteors and that as these crashed into the Sun an immense burst of energy was released. But there was no evidence for such a supply of meteors, nor for the Sun growing in mass as it absorbed all the spent meteors. In 1862, the Irish physicist William Thomson (later Lord Kelvin) adapted the theory to suggest that the Sun's energy came from the early agglomeration of much smaller bodies in the solar system. As they were drawn together and crushed into the form of the Sun by gravity, becoming ever denser, massive amounts of energy were generated. The Sun was, then, 'an incandescent liquid now losing heat'. This drew on a proposal by the German physicist Hermann von Helmholtz that the Sun had begun life as small particles, even dust, which were slowly drawn closer and closer together by gravity. But Thomson's calculations showed the Sun could not sustain its output of energy for

> 'It seems, therefore, on the whole most probable that the sun has not illuminated the earth for 100,000,000 years, and almost certain that he has not done so for 500,000,000 years. As for the future, we may say, with equal certainty, that inhabitants of the Earth can not continue to enjoy the light and heat essential to their life for many million years longer unless sources now unknown to us are prepared in the great storehouse of creation.'
>
> William Thomson,
> 'On the Age of the Sun's Heat', 1862

more than about 20 million years at the rate of solar energy output estimated by French physicist Claude Pouillet, though he allowed that greater density at the centre might push this up to 100 million years. Thomson then based his estimate for the age of the Sun on his theory of how long its energy supply could last. His final estimate, issued in 1897, was 20–40 million years, and probably closer to 20 million. Even at the time, geology and evolutionary biology suggested a much greater age. In 1895, Thomson's erstwhile assistant John Perry published a paper suggesting the age of the Earth was 2–3 billion years.

Thomson's argument soon suffered another blow. Radioactivity was discovered by the French scientists Henri Becquerel and Marie Curie, and the method of radiometric dating developed in the first years of the 20th century. By 1911, a sample of rock had been dated at 1.6 billion years, and an age of 4.55 billion ±0.07 million years was established in 1956.

Radioactivity itself then became a new candidate for the power source of the Sun. A calculation in 1903 suggested that if there were just 3.6 grams of radium for every cubic metre of the Sun, its decay would produce enough energy to match the Sun's output.

The real breakthrough came in 1905 with Albert Einstein's famous equation $E=mc^2$ which equates energy and matter. Clearly, the energy for the Sun could come from atoms.

William Thomson, later Lord Kelvin, was one of the greatest physicists of the 19th century, but doggedly adhered to his belief that the Sun is relatively young.

Introducing . . . helium!

The question of how the Sun works could not be solved without knowing what it is made of. The answer to that question came with the development of spectroscopy. Fraunhofer had discovered spectroscopy in 1814 by looking at light from the Sun and finding the spectrum of hydrogen. Although interesting, that didn't help immediately to solve the problem.

Another clue came in 1868 when two astronomers working independently, Norman Lockyer in England and Pierre Janssen in France, discovered a yellow line in the spectrum of the Sun that could not be matched with any known element. Lockyer boldly – and correctly – proposed that it represented a new element not found on Earth. He named it helium, from the Greek *Helios*, god of the Sun.

The fuel revealed

Cecilia Payne-Gaposchkin (1900–79), a young Englishwoman, was the first student to earn a PhD from the Harvard Observatory. She did it with a thesis of unparalleled brilliance. *Stellar Atmospheres* revealed that although the Sun contains elements similar to those found on Earth, plus helium, there is a considerable difference in their proportions. She declared that most of the Sun's atmosphere is hydrogen; it is, she said, a million times more abundant than the other elements. The leading American astronomer of the day, Henry Norris Russell (1877–1957), disagreed, saying it was 'clearly impossible', and he persuaded Payne-Gaposchkin to accept that her findings were wrong. But she was soon vindicated. Not long after, Russell came to the same conclusion, publishing his findings

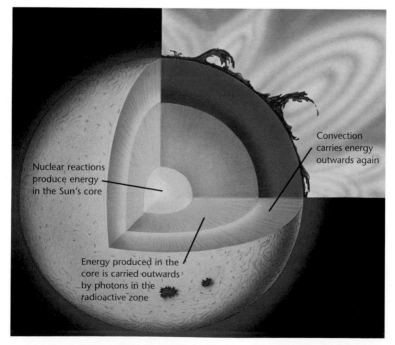

Nuclear reactions produce energy in the Sun's core

Convection carries energy outwards again

Energy produced in the core is carried outwards by photons in the radioactive zone

Energy produced at the Sun's core follows a long and circuitous path to the surface because of the density of the Sun. It might take anywhere between 4,000 and a million years for a photon to travel from the core to the surface.

in 1929, with full credit to Payne-Gaposchkin. With his backing, the notion that the stars are made largely of hydrogen was soon accepted. That laid the foundations for discovering how the Sun, and all other stars, work.

Arthur Eddington, Payne-Gaposchkin's early mentor, proposed in 1927 that the Sun might be powered by nuclear fusion, with the

Cecilia Payne-Gaposchkin worked out the composition of stars, but her conclusion was initially dismissed as 'impossible'.

intense temperatures at its heart forcing hydrogen molecules to fuse together to form helium. Eddington was an expert on Einstein's theory of relativity and well aware that energy and matter are ultimately interchangeable. The same year, working independently, Jean Baptiste Perrin came up with the same idea in France. It would not be verified until 1939.

Hans Bethe, a German scientist with a Jewish mother, had left Germany with the rise of the Nazis and gone to the USA where he later worked on the Manhattan Project, the secret programme to build a hydrogen bomb during World War II. Bethe saw the connection between the project and the Sun. He worked out that at the heart of the Sun, temperatures are so high that atoms disintegrate, losing their electrons. The nuclei of hydrogen atoms (each containing one neutron and one proton) travel in pairs as hydrogen forms diatomic molecules. The immense heat and pressure force them

closer together with such force that some become welded, forming helium nuclei. (Helium has two protons and two neutrons in the nucleus.) The resulting helium nucleus has a marginally smaller mass than the hydrogen nuclei: 0.7 per cent of the mass is converted into energy, which is released. This energy powers the Sun. This now forms the Standard Solar Model. Every star works in the same way.

STILL GOING STRONG

The Sun converts four million tonnes of its mass to energy every year – and has done so for the past five billion years. There's still plenty left: it has used up only a thousandth of its mass in this way.

CHAPTER 5

The solar system
REVEALED

'The Sun strings these worlds – the Earth, the planets, the atmosphere – to himself on a thread.'

Yajnavalkya,
Vedic philosopher,
7th century BC

The five planets visible to the naked eye – Mercury, Venus, Mars, Jupiter and Saturn – have been known since prehistoric times. Early astronomical records reveal that some difference between the planets and the stars was already known. Yet the extent to which our solar system is a unit, and the way in which it relates to the rest of the universe, has only emerged over the last 400 years.

Our solar system: the rocky, gassy bodies of the planets and their myriad moons are still revealing their secrets.

Exploring the planets

Observations of Venus recorded in the 7th century BC, but probably dating from the 2nd millennium BC, show that Babylonian and probably Sumerian astronomers were aware that the planets, or 'wandering stars', moved differently from the 'fixed stars'. The Aristotelian and Ptolemaic account of the heavens placed the fixed stars on the outermost sphere, beyond the planets. It essentially considered the Moon, Sun, and the five planets visible to the naked eye in much the same way, with all orbiting the Earth on a displaced circle, or 'deferent'. With the Copernican model, the Earth took its place as a planet and the Sun and Moon relinquished their places as *de facto* planets. The solar system was defined and, with the advent of the telescope, open for business.

Even without a telescope, the ancients detected the reddish hue of Mars, but no other details of the planets could be seen. The planets beyond Saturn were not visible at all, and they remained hidden, even when Galileo turned his first telescope towards the night sky.

Moons galore

Galileo's first great discoveries related to the Moon (see page 104), but he soon discovered that the Moon was not the only natural satellite in the solar system. When he turned his telescope towards Jupiter, he saw three dim stars alongside it. What was remarkable was that on subsequent nights, the 'dim stars' were in different places but still near Jupiter. He found that after a while one of them disappeared, then it reappeared, and finally a fourth became visible. A large part of *The Starry Messenger* (see page 80) is taken up with plotting the vagaries of the mysterious bodies, which he named the Medicean stars in order to curry favour with Cosimo de' Medici. He noticed that the stars moved with Jupiter against the background of the fixed stars, and that they varied in size, night by night.

The Romans represented the five planets as gods. Here Mercury is shown entrusting the child Bacchus to the care of nymphs.

Galileo sketched the arrangements of Jupiter and its moons that he observed.

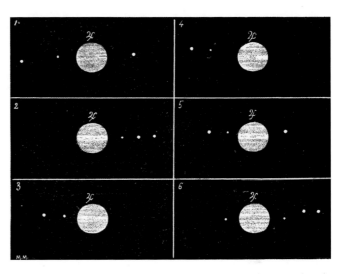

Galileo concluded, correctly, that these were bodies orbiting Jupiter. They are Ganymede, Io, Europa and Calisto, the four largest of Jupiter's moons, now known as the Galilean moons. He noticed that their movements did not form a consistent pattern, which he correctly attributed to each moon having a different orbit. He was incorrect in thinking that the variation in the apparent sizes of the moons was because Jupiter had a layer of atmosphere (a 'vaporous sphere') around it. He thought that this made the planets look fainter when seen through it and brighter when in front of it. He suggested that all planets in the solar system (and the Moon) might have an atmosphere, but made no more of it. Perhaps the fact that an atmosphere might suggest the possibility of life was a step too far along the road to heresy, especially coming only ten years after Giordano Bruno of the 'infinite worlds' theory had been burned at the stake (see page 194); caution was understandable.

Saturn's problematic ears

Turning to Saturn, Galileo found that this planet presented its own puzzle. He noted 'ear-like' projections on its sides.

GANYMEDE SPOTTED?

The Chinese astronomer Gan De made detailed observations of Jupiter and in 365BC reported that he had seen a small star with a reddish hue alongside Jupiter. The Chinese historian of astronomy Xi Zezong (1927–2008) argued that this was a naked-eye sighting of Ganymede. However, there is no reason for Ganymede to appear red, as its surface is made of silicate rock and water ice.

'I discovered another very strange wonder, which I should like to make known . . . keeping it secret, however, until the time when my work is published . . . the star of Saturn is not a single star, but is a composite of three, which almost touch each other, never change or move relative to each other, and are arranged in a row along the zodiac, the middle one being three times larger than the lateral ones, and they are situated in this form: oOo.'

Galileo, 1610

Saturn's rings are very thin, so when the planet is appropriately tilted, as seen from Earth, they disappear – a disappearance that greatly troubled Galileo.

Subsequent observers saw the planet as oval, but Galileo put this down to using inferior telescopes. It was troubling that the satellites, if that was what they were, were so much larger in relation to the planet than were the satellites of Jupiter. More

ANAGRAMS FOR PRIORITY

Before the days of establishing priority by publishing a paper in a journal, scientists needed different methods of laying claim to a discovery during the long period before a book might be published. One method was to issue an anagram that contained the coded discovery. The anagram could be revealed if anyone else happened to come up with the same discovery before publication. Galileo's anagram for the discovery of Saturn's odd shape was: 's m a i s m r m i l m e p o e t a l e u m i b u n e n u g t t a u i r a s'.

The resolved anagram was *Altissimum planetam tergeminum observavi*, or 'I have observed the highest planet tri-form.' With 37 letters, it could not easily be reconfigured into anything else to lay claim to another scientist's discovery – though Kepler did believe he had solved this anagram, rearranged to *Salue umbistineum geminatum Martia proles* ('Hail, twin companionship, children of Mars'), to report the discovery of two moons around Mars! The Dutch mathematician Christiaan Huygens produced an anagram encoding his own theory about Saturn in 1656, revealing it in full in his book of 1659 (see opposite).

Huygens illustrated the different appearances of Saturn in his Systema Saturnium *(1659) and outlined his theory of a thick, solid ring girdling the planet.*

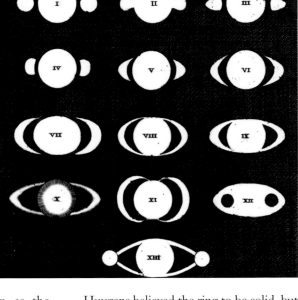

troubling still for Galileo was his observation in 1612 that the 'ears' had disappeared.

In 1656, Johannes Hevelius proposed that Saturn is an ellipsoid with crescents on either side. Rotation about the minor axis in the plane of the crescents would explain all the planet's appearances. Not many people took this explanation seriously. Then, in 1658, Christopher Wren (better known as the architect of St Paul's Cathedral in London) suggested that there was a very thin corona around the planet, and it orbited around the major axis. The problem was essentially solved by Huygens and revealed in 1659: he suggested that there is a thin, flat ring surrounding, but not touching, the planet.

Huygens believed the ring to be solid, but in other regards his solution was accurate. Jean Chapelain (1595–1674) suggested in 1660 that the ring was made up of many tiny chunks or very small satellites, but he was better known as a critic and poet and was largely ignored. Most astronomers remained convinced that the ring was solid. The Italian astronomer Giovanni Cassini argued in 1675 that Saturn has numerous rings with gaps between them. In 1858, the physicist James Clerk Maxwell showed mathematically that the ring must be composed of chunks no larger than a

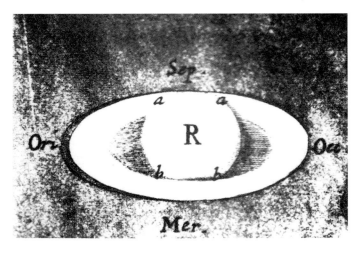

Robert Hooke's drawing of Saturn from 1666 clearly shows the separation of ring and planet.

GIOVANNI CASSINI (1625–1712)

After an early interest in mathematics and astrology, Cassini turned to astronomy, working at the Panzano Observatory from 1648 to 1669 and as professor of astronomy at the University of Bologna. In 1669, he moved to Paris to set up the observatory there. With the English scientist Robert Hooke, Cassini is credited with the discovery of the great spot on Jupiter around 1665. He discovered four of Saturn's moons and the gap in Saturn's rings, now known as the Cassini Division.

In 1672, Cassini stayed in Paris while his colleague Jean Richer went to Cayenne, French Guiana. From their respective positions, the pair observed Mars simultaneously and used parallax to calculate its distance from Earth. This was the first calculation of a distance between planets in the solar system using a telescope.

Cassini was also one of the first people to measure longitude successfully. He put it to use in an ambitious project to measure France accurately, but came up with the politically disappointing result that the country was much smaller than previously believed.

---

few inches across. In fact, the particles in the rings vary from just microns across to the size of mountains. They are mostly ice, with some rock dust, and there is much empty space between the solid matter.

Jupiter – spots, bands and storms

Giovanni Cassini was also the first to notice the great red spot on Jupiter, drawing the spots and bands of the planet in the 1660s. It is on the basis of Cassini's account that most current astronomers believe that the great red spot is a storm that has been raging for at least 350 years. But when it first appeared or even whether the same storm has been raging continuously since Cassini's time are unknown.

How many planets?

While 17th-century astronomers discovered the moons of Jupiter and Saturn, they did not add any new planets to the solar system. This would only change with a discovery in the late 18th century which came as a surprise to everyone, including the astronomer who made it.

CASSINI GOES TO SATURN

Launched in 1997, the Cassini–Huygens mission to explore Saturn, together with its rings and moons, was named after two astronomers, Giovanni Cassini and Christian Huygens. It was made up of an orbiter spacecraft, Cassini, and a detachable lander, Huygens. The lander touched down on Titan, revealing the methane lakes that cover much of that moon's surface. The Cassini orbiter found that Saturn's rings are typically only about 30m (98ft) thick, which explains why they become invisible when viewed edge-on from Earth. The mission, which is managed by NASA, is a collaborative effort involving 17 countries.

The red spot of Jupiter, clearly visible in this photo from NASA, represents a storm that has probably raged since at least the time of Cassini and possibly for hundreds or thousands of years longer.

Not a comet

William Herschel was an enthusiastic builder of his own telescopes (see page 84). In 1781, he was hunting for double stars – pairs of stars so close together they look like one bright star unless sufficiently magnified – when he found what appeared to be a moving star. He assumed he had found a comet and sought it out on subsequent nights. But the object travelled too slowly for a comet. Furthermore, instead of being a point of a light, it was a fuzzy, light disc. For a comet to be so bright it would need to be near the Sun, yet it moved too slowly for this to be the case. From this Herschel deduced that he had found a new planet – the first to be discovered since prehistoric times.

129

The icy planet Uranus has a very faint ring system, discovered in 1977 – but Herschel reported seeing rings in 1789. Some astronomers think it unlikely he could actually have seen them.

He wanted to name the new planet George in honour of the king, George III, but this was unpopular with other European astronomers and it became known instead as Uranus. Pleased that this great discovery had been made on British soil, the king awarded Herschel a generous pension in recognition of his discovery. This allowed Herschel to give up his day job as a musician and focus on astronomy. He was assisted by his sister Caroline, who would become an accomplished astronomer in her own right. It also allowed him to build more and bigger telescopes. He and Caroline went on to discover more moons and compile a catalogue of 2,500 stars.

In fact, Uranus was probably observed in 1690, when it was thought to be a star in the constellation of Taurus, but no astronomer before Herschel had a telescope powerful enough to resolve the 'star' into the disc of a planet. When Herschel sought confirmation of his discovery from other astronomers, none of them could see what he saw, as their telescopes were inferior to his.

It's hard to imagine the impact of a new planet in the 18th century. Although science was no longer hamstrung by the Church's insistence on the unchanging nature of the universe, the addition of Uranus meant the solar system had suddenly doubled in size.

The missing planet

In 1766, German astronomer Johann Titius showed that the spacing of the planets follows an approximate, predictable pattern, with the exception of a gap between Mars and Jupiter, where a planet should be located but is missing. Kepler had already noticed this gap. A few years later, in 1778, Johann Bode formulated the relationship as a rough mathematical expression (known as Bode's Law, or the Titius-Bode Law) and predicted that a planet should be found in the gap.

In 1800, the Hungarian astronomer Baron Franz Xaver von Zach formed a club with 24 other astronomers called the United Astronomical Society (or sometimes the 'Stellar Police'). William Herschel was a member, and an Italian monk, Giuseppe Piazzi (1746–1826), had been invited to join. Before accepting (or perhaps even receiving) his invitation, Piazzi seemed to have discovered the missing planet. In exactly the predicted orbit, he found an object that looked like a star but moved like a planet. No matter how much it was magnified, though, it never resolved into a disc. Piazzi observed the object for 41 days, but then fell ill and was unable to continue. By the time he was well again, it had moved too close to the Sun to be visible.

Finding this missing planet proved a challenge to contemporary mathematicians. French astronomer Pierre-Simon Laplace (1749–1827) said it was impossible. Yet in December 1801 a German mathematician, relatively unknown at the time, made a name for himself by computing the orbit and predicting accurately where the body would be found. He was Carl Friedrich Gauss (1777–1855), only 24 years old at the time. Gauss, who later became a famous mathematician and astronomer too, never fully revealed his method. The object was named Ceres. Just 15 months later, another member of the society, Heinrich Olbers, found a second, almost identical object in a very similar orbit. Herschel suggested that they should be called 'asteroids', from the Greek 'star-like'. Two more turned up in 1807 and another in 1845. The pace picked up and soon asteroids were appearing thick and fast until by the middle of the century there were 23 planets. It was clearly time for a rethink. As Herschel had coined the term 'asteroid' in 1802, Ceres and its companions were reclassified as asteroids, leaving only the established seven planets, including Earth. But no one thought to come up with a formal definition for a planet, leaving the door open for problems in future.

The asteroid count continued to mount, reaching the hundred mark by 1868, increasing to 1,000 by 1921, and to more than 100,000 by 2000. There are now known to be millions, ranging in size from dwarf planets to microscopic particles.

Olbers suggested to Herschel that the asteroid belt (as it is now known) might have been formed by the destruction of a large planet in the empty orbital slot between Mars and Jupiter. Popular for a while, the theory is now largely rejected. The composition of all the asteroids is not the same, as would be expected from the destruction of a single planet. And although there are lots of asteroids, there is also a lot of empty space in the asteroid belt. The estimated total mass of all the asteroid material comes to only 4 per cent of the

The asteroids orbit the Sun, mostly in a broad belt between Mars and Jupiter.

mass of the Moon, far too little to have made a planet. It's currently thought that the asteroid belt represents a 'missed planet' that failed to form from the protoplanetary disc at the start of the solar system. It may have been prevented from joining up by the huge pull of Jupiter's gravity. In that case, the asteroid belt is a useful relic of the early days of the solar system.

More planet-hunting

Soon after Uranus was discovered, abnormalities in its orbit became apparent. In 1808, French astronomer and mathematician Alexis Bouvard (1767–1843) published tables detailing the orbits of Jupiter and Saturn. As Uranus has an 84-year orbit and had been discovered only 27 years previously, there was not enough data for Bouvard to include it in

his tables, but he was keen to have it in his next publication. Bouvard decided to look at historic sightings of bodies that could retrospectively be recognized as Uranus. Using these, he forecast its orbit – but the planet did not act as expected over the coming years. Bouvard proposed the presence of another planet exerting a gravitational force on Uranus, but he was unable to secure help in finding it.

In 1846, just three years after Bouvard's death, the expected planet was found and named Neptune. The French astronomer Urbain Le Verrier and the Briton John Couch Adams independently predicted the position of the additional planet working from Bouvard's tables. Le Verrier had no luck interesting French astronomers in the search and sent his data to the Berlin Observatory, where Johann Gottfried Galle found Neptune within just one degree of the predicted location on his first night looking for it. The planet was 12 degrees away from the position predicted by Adams. A dispute over precedence was settled with both Adams and Le Verrier being credited equally with the discovery. Neptune's largest moon, Triton, was found just 17 days after the planet itself. Neptune's 165-year orbital period means that only one 'Neptune-year' has passed since its discovery.

Looking more closely – the Red Planet

Mars had been observed since prehistoric times, and its red hue was well known. Tycho Brahe made regular observations of Mars as he struggled to plot its precise orbit.

Neptune, another icy planet and the outermost planet known, has active weather systems which manifest as swirls and spots.

LIFE ON MARS?

Mars has seemed the best prospect for life outside Earth since Herschel's early discoveries (see opposite). The American amateur astronomer Percival Lowell became convinced there were canals on Mars which had been built by intelligent beings, an idea he promoted enthusiastically. It sparked the imagination of many writers, including H.G. Wells, whose *War of the Worlds* (1898) imagines an invasion of Earth by beings from Mars. Mars remained the favoured source of aliens for science fiction for most of the 20th century, as well as the best bet for finding other life in the solar system. But probes sent to Mars in the 1960s and 1970s revealed a barren landscape of rock. The dream of abundant and intelligent life there was quashed.

Currently, NASA says Mars has had liquid water in the past and may have liquid water beneath the surface, but the planet is too cold for surface water at present. Microscopic life might have existed in the past; current and future missions will try to establish evidence for it.

Kepler used observations approximately one Martian year apart as the basis of his parallax calculations, finding the distance from Mars to the Sun to be 1.5 times the distance from Earth to the Sun.

Mars has no moons or rings, and nothing interesting could be seen with the earliest telescopes. The planet only started to reveal its secrets in the middle of the 17th century. Huygens drew a sketch of Mars in or before 1659, but it was no more than a scribble of shadow on a circle. Even so, he saw the dark patch consistently enough

to be able to assess the rotational period of Mars, declaring it to be 24 hours. In 1666, Cassini noticed the polar ice caps on Mars without realizing what they were. But it was in the late 18th century that Mars proved really interesting. Herschel

Mars, the rocky planet most similar to Earth and often our nearest neighbour, looks red because of the preponderance of iron, as rust, on its surface.

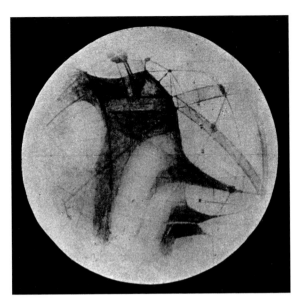

A drawing of Mars, with 'canals' and dark areas, that Percival Lowell made in 1896. He believed Martians had built the canals to carry water from the poles to irrigate their land.

observed Mars extensively and drew some interesting conclusions. First, Mars is tilted on its axis, like the Earth, so has seasons during the course of its year, though its year is around twice that of Earth's. The ice caps grow and shrink with the seasons, so he suggested that they might, like Earth's ice caps, be made of snow and ice. Herschel calculated the length of a day on Mars as 24 hours, 39 minutes and 22 seconds, just 14 seconds shorter than the currently accepted figure. Herschel believed he saw clouds move over the surface of Mars and suggested that the shifting dark and light areas might be seasonally flourishing vegetation. The idea that Mars might have life slowly emerged.

The orbit of Mars brings it especially close to Earth every 15–17 years. It was during one of these periods, in 1877, that the American astronomer Asaph Hall (1829–1907) spotted Mars's two small moons and named them Phobos and Deimos. In the same year, the Italian astronomer Giovanni Schiaparelli (1835–1910) made the first detailed map of Mars. Schiaparelli described lines across the surface of Mars, which he designated *canali* and gave them the names of rivers on Earth. The lines were later shown to be an optical effect rather than a feature of the planet, but a ball had been set rolling. The mistranslation of *canali* as

'canals' (rather than channels or grooves) excited many people, including a wealthy American businessman with a keen interest in astronomy. Percival Lowell (1855–1916) calculated that Mars would be close to Earth in 1894 and decided to build an observatory to view the surface of Mars clearly. The Lowell Observatory in Flagstaff, Arizona, opened in 1894 in time for Lowell's Mars observations and he drew his own maps of the 'canals' of Mars. The idea of life on Mars became hugely popular, sparking science fiction stories and establishing a motif that continued throughout the 20th century. Yet the development of better telescopes soon removed the artefact of the *canali* – even by 1909 the French astronomer Camille Flammarion could see patterns on Mars, but no canals.

Planet X, or O, or Hyperion

Neptune accounted for part of the discrepancy in Uranus's orbit, but the orbit was still not quite as expected. Several astronomers suggested that an even more distant planet still lurked in the darkness, but they showed more skill at naming the unknown planet than finding it. The French physicist Jacques Babinet proposed naming it Hyperion in 1848; his countryman Gabriel Dallet called it Planet X in 1892; but the American astronomer William Henry Pickering preferred Planet O. The hunt for the mystery planet was on.

After Lowell had drawn his images of Mars in 1894 and the Red Planet had moved on, he was left with an idle observatory. He switched over to looking for the mystery 'Planet X', employing a team of astronomers who spent years on the search. It was carried out by taking photographs of an area of sky, night after night, and comparing them to find anything that might have moved. Lowell died of a brain haemorrhage in 1916, but the search continued.

When Lowell died, the man who would eventually find 'Planet X' was only ten years old. Clyde Tombaugh (1906–97) grew up on farms in Illinois and Kansas. He had no education in astronomy, but began building his own telescopes in 1926. He dug a pit in which to house them, which doubled as a root cellar and shelter for his family. In 1928, Tombaugh built a telescope from the crankshaft of a 1910 Buick and parts from a cream separator, also grinding his own mirrors for the reflector. He used it to

observe Jupiter and Mars, making drawings of what he saw. He sent his drawings to the Lowell Observatory hoping to get some professional feedback. Instead, he got a job.

At just the time Tombaugh sent in his drawings, the Observatory was looking for someone to work using a 'blink comparator', a device that flipped rapidly between two photographs allowing easy comparison. Any movement of objects in the foreground was, in theory, easy to spot, but as there were up to a million stars in each photo, it wasn't that easy. Tombaugh joined the project in 1929 and found the planet early the following year. A worldwide competition to name 'Planet X' was won by an 11-year-old English girl, Venetia Burney, who proposed the name Pluto after the Greek god of the

Clyde Tombaugh with his homemade telescope in 1928.

Venetia Burney, who named Pluto at the age of 11.

in the asteroid belt, they were beyond the orbit of Neptune. The first was Sedna, found in 2003 by American astronomer Michael Brown (b.1965). As bodies of similar size and with similar orbits followed, it became clear that the skies could soon be crowded with small planets again. In 2006, the International Astronomical Union (IAU) agreed a formal scientific definition for a 'planet' for the first time and Pluto lost its planet badge. A planet, the IAU ruled, must orbit the Sun; be large enough to have become round under the force of its own gravity; and dominate the area around its orbit. Pluto passes the first two criteria, but it has allowed its orbit to remain untidily cluttered with asteroids and other debris which any self-respecting planet would have swept up into its own body. One of Pluto's moons, Charon, is about half the size of Pluto itself, which also violates the behavioural standards expected of planets.

underworld. She was rewarded with £5. Tombaugh, who died in 1997, had a rather special (if belated) reward. In 2006, his ashes were carried to Pluto by the NASA New Horizons mission, a space probe sent to study the planet Tombaugh had discovered.

The unplaneting of Pluto

Pluto's reign as a planet was relatively short-lived. Just as the 1800s had seen a rush of asteroids-that-might-be-planets, the 21st century saw a further flurry of potential planets. This time, instead of being

Pluto is now considered to be one of many dwarf planets or large Kuiper objects in the Kuiper belt beyond the orbit of Neptune.

PLANET X – REPRISE

The original Planet X eventually materialized as Pluto, but the name Planet X is in use again. It sounds like an alien world from a cheap sci-fi film, and the possibly-there planet has taken on a life that could fit well in that genre. Many astronomers prefer the name 'Planet Nine' as it's less evocative of loony-fringe theories. There is both a mainstream suggestion that there might be another planet beyond the orbit of Pluto and a trend in speculative astronomy to 'find' a planet apparently known to ancient civilizations, but since lost.

In theory, it is possible that a planet or dwarf planet with a widely eccentric orbit could have been close enough to Earth to have been seen thousands of years ago, but has since moved too far away to be visible. It would need a very eccentric orbit, taking thousands of years to complete, but it's not impossible: the dwarf planet Sedna has a highly elliptical orbit that takes 11,600 years to complete.

In 1976, Russian-American author Zecharia Sitchin started to popularize the presence of an additional planet among pseudo-astronomers. Sitchin claims the planet is documented in ancient cosmological texts from Mesopotamia and is called Nibiru. According to ancient wisdom, he says, the planet is on a highly elliptical orbit that brings it close to Earth every 3,600 years. So far, so good – just about. But he went on to say that its return is associated with disastrous events including earthquakes, tsunami, mass extinction and collisions with comets. The planet's most recent predicted return (though not by Sitchin) was in 2003, but both planet and disasters went unnoticed. Further, Sitchin's interpretations of cuneiform texts and a seal he claims shows Nibiru as a planet are rejected by astronomers and experts in the languages involved. Nevertheless, the possibility that there is at least one further planet at the outer reaches of the solar system remains.

The orbits of the four outer planets (Jupiter, Saturn, Uranus and Neptune, shown in green), three dwarf planets (Ceres, Pluto and Eris – yellow) and ten candidate dwarf planets (brown).

The Leonid meteor shower was observed by the French aeronauts Henri Giffard and Wilfrid de Fonvielle from a hot-air balloon in 1870.

So Pluto was de-planeted and has become a dwarf planet, designation 134340 Pluto, just one of many large objects in the Kuiper belt. Other dwarf planets include Ceres in the asteroid belt and Eris, which lies beyond Pluto's orbit.

Rocks from space

Meteors are very short lived. Sometimes called shooting stars, they appear as a bright point of light that comes from nowhere, streaks briefly across the sky and then vanishes. Before they enter Earth's atmosphere they are known as meteoroids – chunks of rock flying through space. They burn up as they are heated by the friction caused by passing through Earth's atmosphere. Most vaporize completely, but if any part survives and lands it is called a meteorite. Most meteorites are tiny, but they can weigh many kilograms. Very large meteoroids are known as asteroids. They are generally chunks of rock from the asteroid belt that orbits the Sun between Mars and Jupiter (see page 131), but some are lumps of rock blasted into space when meteorites collide with the surface of the Moon or Mars. Meteorites that originate from Mars are particularly important, giving clues to the early development of the Red Planet.

The Greek philosopher Anaxagoras suggested in the 5th century BC that meteors were stars (which he thought were burning lumps of rock and iron) that had been shaken

'On the night of November 12–13, 1833, a tempest of falling stars broke over the Earth. . . .
The sky was scored in every direction with shining tracks and illuminated with majestic fireballs.
At Boston, the frequency of meteors was estimated to be about half that of flakes of snow in
an average snowstorm. Their numbers . . . were quite beyond counting; but as it waned, a
reckoning was attempted, from which it was computed, on the basis of that much-diminished
rate, that 240,000 must have been visible during the nine hours they continued to fall.'

Astronomer Agnes Clerke, describing the 1833 Leonids meteor shower

out of place and fallen to Earth. Apart from them not being stars, he wasn't far wrong. But his view was soon lost and for many centuries meteors were considered not to be from space at all but to be an atmospheric phenomenon, like lightning. This fitted the Aristotelian model in which only the sublunar region is subject to change and only objects in that region are capable of rectilinear motion.

Meteors often come in showers, which occur at regular times during the year when Earth passes through a cloud of debris left by a disintegrating comet. It was one such spectacular shower that reignited interest in meteors and led to them being identified as cosmic in origin.

The Leonids form a show of meteors visible in November each year. They get their name from the fact that they seem to originate from the constellation Leo. The first reported sightings were in AD902 when they were recorded in China, Italy and Egypt. A particularly spectacular display in 1833 led many terrified Americans to believe that the Final Judgment was at hand. This

The Leonids made an exceptionally spectacular display in 1833, shown here over North America.

The Chelyabinsk meteor exploded in Earth's atmosphere over the Russian city in 2013 around 30km (18½ miles) above the ground .

event has even been held responsible for a renewed religious fervor at that time and the establishment of new religious sects that still exist today.

The spectacular display of the Leonids caused astronomers to think again about meteors and to examine ancient records of meteor showers, not just from Europe but in Chinese and Arab records, too. In 1837, the German astronomer Heinrich Olbers suggested that the most dazzling displays appeared every 33 or 34 years. Soon, astronomers traced the meteors to a 'knot' of matter orbiting the Sun on that timescale. In 1866, the return period was established as 33.25 years. This is the return period of the comet responsible for the Leonids; the 'knot' of matter is replenished each time the comet returns. The same year, Giovanni Schiaparelli, famous for the 'canals' on Mars, identified the link between the Perseids meteor shower (seen in August and named after the Perseus constellation) and the comet Swift-Tuttle. Links between other meteor showers and comets soon followed, including that between the Leonids and a small comet called Tempel-Tuttle.

Visitors with a bad reputation

Meteors are short-lived, over in a moment, although many may appear together. This is not the case with comets, often considered to be portents of evil or disaster. Comets

John Everett Millais' depiction of the subjugation of the Inca people, defeated by Spanish conquistadors in 1532. The Incas retrospectively saw the appearance of Halley's comet in 1531 as a harbinger of doom.

were blamed for all types of terrible events, including the Black Death (Comet Negra, 1347) and the slaughter of the Incas by Pizarro's invading troops in 1532 (Halley's comet, 1531). An extraordinary coincidence of two comets and a lunar eclipse in 1664–5 led English astrologer John Gadbury to warn in 1665 that: 'These Blazeing Starrs! Threaten the World with Famine, Plague, & Warrs. To Princes, Death: to Kingdoms, many Crises: to all Estates, inevitable Losses!' Luckily for him – if not for anyone else – 1665 saw the Great Fire of London and 1666 the Great Plague, vindicating his prophecy of doom. But, of course, the comets were seen everywhere and not everywhere suffered a catastrophe such as a major conflagration and plague.

Comet-phobia continued long after the nature of comets was understood. When Halley's comet was due to return in 1910, the Western world was galvanized by terror. Spectroscopy had made the comet even scarier than before, as the Yerkes Observatory announced that the tail contained the deadly gas cyanogen and would swamp Earth and possibly destroy all life on the planet (though clearly it hadn't done on its previous visits). People

BAD PRESS FOR ANOTHER COMET

In 1997, 39 members of the American religious cult Heaven's Gate killed themselves in the belief that they would then be picked up and elevated to a new level of existence by an alien spacecraft trailing the comet Hale-Bopp. In 1996, the leader of the cult had taken out insurance against alien abduction for 50 of its members.

were duped into buying comet masks so that they could breathe safely, and comet pills to counteract the effects of the gas.

Although it's easy to mock this scaremongering, a direct hit by a comet could be disastrous for Earth. Scientists believe that the extinction event 65 million years ago which wiped out the non-avian dinosaurs was precipitated by a comet or asteroid colliding with Earth in what is now the Gulf of Mexico. Other extinction events might also have been triggered by such occurrences. Since 1998, NASA has been tracking near-Earth objects and has so far discovered 13,500 of them.

'Broom' stars

A comet's most obvious characteristic is its tail, which clearly distinguishes it from planets and stars. It is the tail that gave them their name: 'comet' is from the Greek word for 'hairy'. The Chinese called them 'broom stars' or 'pheasant-tailed stars'.

Chinese astronomers have kept records of comets for more than 2,000 years, and these records have proved invaluable to modern astronomers working out the return periods of comets.

Until the late 16th century, it was believed that comets, like meteors, were close to Earth – lower than the orbit of

WHERE COMETS COME FROM

Astronomers divide comets into three kinds: periodic or short-period comets, long-period comets and single-apparition comets.

Short-period comets orbit the Sun in periods of 200 years or less, so are visible at least every 200 years. Halley's Comet is a famous short-period comet with a return period of around 75–6 years. Short-period comets are believed to originate in the Kuiper Belt, a region beyond the orbit of Neptune.

Long-period comets take more than 200 years to orbit the Sun, and come from the Oort Cloud on the outer regions of the solar system. Hale-Bopp was a very bright long-period comet seen in 1996–7, its first appearance in 4,200 years. Long-period comets have a highly eccentric (elliptical) orbit; they loop closely around the Sun and then head off towards the outer reaches of the solar system, spending most of their orbit a very long way away.

Like long-period comets, single-apparition comets originate in the outer reaches of the solar system. Their path takes them close to the gas giants, whose gravity affects their orbit just enough to send them out of the solar system, never to return.

COMET CAESAR

The brightest comet in recorded history was Comet Caesar (C/–43 K1) which was seen in 44BC and interpreted in Ancient Rome as a sign that the recently assassinated Julius Caesar had become a god. The Roman historian Galius Suetonius recorded that 'a comet shone for seven successive days, rising about the eleventh hour, and was believed to be the soul of Caesar.' It was visible even in daytime. It's likely that the comet was non-period and might have disintegrated.

the Moon in the Ptolemaic model of the universe. But after observing a bright comet in 1577, Tycho Brahe calculated that they lie among the planets. This was a major blow to the Ptolemaic model, as the celestial spheres beyond Earth were thought to be unchanging. Brahe's comet (officially designated C/1577 V1) was one of the five brightest comets in recorded history. It has an unknown return period, and may never return; it is currently estimated to be 320 AU from the Sun (Pluto is 40 AU from the Sun).

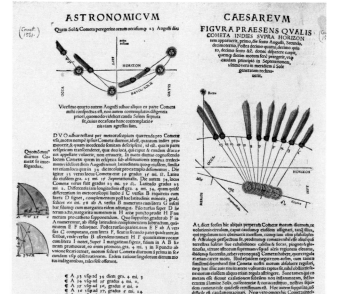

Thinking about comets

A flurry of comets in the 15th and 16th centuries led to renewed scientific interest, as well as pamphlets predicting doom and gloom in their wake. In 1531, the German astronomer Peter Apian recognized that the tail of a comet always

In Astronomicum Caesareum, Peter Apian clearly illustrated comets with their tails always pointing away from the Sun.

Johannes Hevelius illustrated different types of comets in his book Cometographia, *1668.*

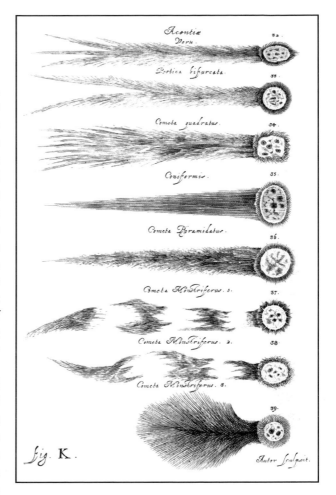

points away from the Sun, though the reason for this was not explained until the 20th century. In 1607, Kepler's observations of a comet led him to conclude that comets travel in a straight line through the solar system (which is untrue) and they are as numerous as fish in the sea but we see relatively few of them (which is pretty much true). He also accurately determined that the tail is produced by parts of the comet evaporating under the influence of the Sun, now known as 'outgassing'.

Johannes Hevelius observed the same comets in 1664–5 that had led Gadbury to his apocalyptic warnings. He came to a more scientific conclusion – that comets originate with planets, particularly Saturn and Jupiter, which hurl them out into the solar system on a trajectory that takes them towards and then curving around the Sun. He did not suggest their path leads to a return visit, though. Still, he had set in motion a train of thought that was on the right track and a few astronomers did begin to wonder whether in fact the paths of comets might be planet-like and constitute an orbit around the Sun in some form.

> 'The direct rays of the Sun strike upon [the comet], penetrate its substance, draw away with them a portion of this matter, and issue thence to form the track of light we call the tail. . . . In this manner the comet is consumed by breathing out its own tail.'
>
> Johannes Kepler, 1607

Halley and the comets

The name most often associated with comets is that of the English astronomer Edmond Halley. His interest was possibly sparked by seeing Tycho Brahe's 1577 comet.

Halley at first struggled to work out the paths of comets. One appeared in 1680, and was seen both before and after its disappearance behind the Sun. But as Kepler's theory that comets travel in straight lines was in vogue, it was not apparent to anyone that it was the same comet, seen twice. Halley failed to work out its motion, unsurprisingly as he was expecting it to travel in a straight line. Another comet appeared in 1682, but although Halley observed it he could not attempt any calculations as he did not have his instruments with him. His later calculations on this comet – which would eventually be named after him – were done from observational data produced by John Flamsteed (see box, opposite).

The true path

In 1684, Halley travelled to Cambridge to consult Isaac Newton on the subject of gravity and found to his surprise that Newton had already worked out his solution to the problem of gravity, but had not yet published it. Halley persuaded him to do so. Newton proposed that comets follow a parabolic path, looping around the Sun and then zooming off into space again. This gave Halley the clue he needed and in 1705, having worked extensively with historical data over ten years, he published his conclusion that comets follow an elliptical orbit through space that brings them close to the Sun and therefore visible from Earth. His table of 24 comets and their apparent

'If according to what we have already said it should return again about the year 1758, candid posterity will not refuse to acknowledge that this was first discovered by an Englishman.'

Edmond Halley, 1749

return periods demonstrated the theory. The comet of 1682, he proposed, was the same comet that Apian had observed in 1531 and Kepler in 1607. (It had also been sighted on many other occasions, the earliest being recorded in China in 240BC.) He suggested the comet would return around 1758 and that, if it did, this would prove his theory. The comet returned as predicted, 16 years after Halley's death, and was subsequently named in his honour.

More excitement

Halley's comet appeared on schedule over the following centuries. Its return in 1910 was surrounded by great public speculation and excitement, with astronomers competing anonymously in a contest to predict its exact course. Despite all the best astronomical minds of the day working on the problem, the comet's predicted closest approach to the Sun (perihelion) was still out by three days. Two astronomers at the Greenwich Observatory, Philip Cowell and Andrew Crommelin, had worked hard to perfect the predictions; they concluded that 'There are forces of an unrecognized kind influencing the comet's motion.' The forces would remain a mystery until 1950 when the American astronomer Fred Whipple (1906– 2004) recognized that the outgassing of a

EDMOND HALLEY (1656–1742)

Edmond Halley was born the son of a wealthy soap-maker in an area that is now the London borough of Hackney, but at the time was a village outside the city. He went to Oxford University in 1673, where he proved to be a brilliant student. Already interested in astronomy, he worked with the Astronomer Royal, John Flamsteed (1646–1719), at the Observatory in Greenwich while still a student. Without finishing his degree, and at the age of only 19, Halley set off to the South Atlantic in 1676 to catalogue the stars of the southern hemisphere while Flamsteed did the same for the northern hemisphere. When Halley returned, having successfully completed the task, he was awarded a Master's degree from Oxford at the command of King Charles II and elected a Fellow of the Royal Society. He was only 22 years old.

Halley then turned his attention to comets, starting with Flamsteed's observations of comet Kirch (1680–1). He attempted to calculate its orbit, but got it seriously wrong, deciding Kirch had an orbital period of 575 years, whereas in fact it is around 10,000 years. After spending ten years making a detailed study of historical accounts of comets going back centuries and consulting his good friend the mathematician and physicist Isaac Newton (see opposite), Halley came to the conclusion that comets follow an elliptical orbit and have a return period that can be calculated mathematically.

Halley was made Astronomer Royal, succeeding Flamsteed, in 1720. His genius spread beyond astronomy: he also invented a working diving bell, published statistics that allowed the more careful calculation of annuities based on life expectancy, studied the Earth's magnetic field and took part in the first attempt to date Stonehenge.

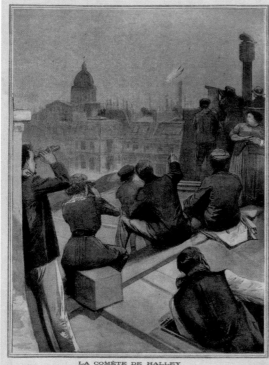

LA COMÈTE DE HALLEY
LES TOITS DE PARIS TRANSFORMÉS EN OBSERVATOIRES

People all over the world turned out to watch Halley's comet in 1910.

images of the impact helped scientists to determine the composition of the comet, showing it to be made of various types of ice and dust. The component particles were found to be finer than expected; around 75 per cent of the comet is empty space.

The Stardust mission, launched in 1999, collected samples of dust from comet Wild-2, which it returned to Earth in 2006. (Dust and gas pour from a comet constantly, so the dust is easily collected without landing on it.) The samples surprised NASA scientists. They were expecting to find mostly tiny grains of 'stardust', very old rocky material formed around previous stars, with a non-crystalline structure. But stardust comprised only a small portion of the comet. Most of

comet acts like rocket propulsion in altering its course. This makes it difficult to predict its orbit solely from the gravitational effects of other bodies.

Up close and personal

More has been discovered about comets since the advent of space travel. There have been several flyby missions to take close-up photographs of comets, and three missions to land on one. NASA's Deep Impact spacecraft crashed into the comet Tempel in 2005. Spectroscopic

The cratered surface of comet Wild-2, photographed by the Stardust spacecraft, shows evidence of bombardment during its past.

the material was made up of larger grains, a lot with a crystalline structure. It seems that comets combine material formed at very high temperatures at the centre of the early solar system (crystals) with material formed at very low temperatures at the edge of the solar system (ices). This has significance beyond the composition of comets, as it yields unexpected information about the formation and distribution of matter in the early solar system.

Going a step further, the European Space Agency's Rosetta mission put the first-ever lander, Philae, on comet C67P/Churyumov-Gerasimenk in 2014. After a bumpy landing, Philae sent back limited data and photos before failing, probably because it landed in shade and couldn't recharge its solar batteries. Rosetta continued to collect data from above C67P and in May 2016 reported finding the amino acid glycine and the element phosphorous on C67P – both essential ingredients for life. Proteins, the chemical building blocks of life, are made up of amino acids. Glycine was suspected on Wild-2; the discovery of an amino acid on two comets suggests that these building blocks might be common in the universe. This fuelled interest in the long-running debate as to whether some of the chemicals needed for life on Earth were delivered by comets billions of years ago and there might be the potential for life elsewhere in the solar system.

The lander Philae on C67P, the first mission to achieve a soft landing on a comet. Philae is about the size of a washing machine.

CHAPTER 6

Mapping the **STARS**

*'A hundred thousand million Stars make
one Galaxy; a hundred thousand million
Galaxies make one Universe.
The figures may not be very trustworthy,
but I think they give a correct impression.'*

Sir Arthur Eddington,
astrophysicist, 1933

**A systematic list that names the stars
and records their position in relation to
one another is called a star catalogue.
For thousands of years, the activity of
astronomers with regard to the stars
focused on counting and cataloguing
them. But as technology improved and
more and more stars were discovered,
making a comprehensive star catalogue
became a formidable task.**

*The patterns of the stars in the sky are easier to remember and see
when turned into pictures, as many cultures have discovered.*

Tracking stars

It's very difficult to describe the position of stars without using some kind of reference aid. There are two obvious solutions: one is to use a coordinate system that allows measurement from a position or line, and the other is to group the stars into named patterns and then identify them by their position in relation to these patterns. The second is the easier method: to see the stars in groups and then use the groups as 'jumping off points' for locating stars. This is the system that emerged, independently, among ancient astronomers the world over. It is a natural human inclination to see patterns in the phenomena around us. Collections of dots or star-shapes that correlate with groups of stars occur even in prehistoric art, suggesting our earliest ancestors also made pictures from the stars.

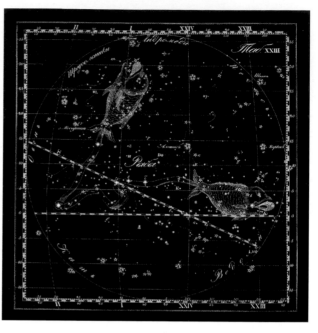

The constellation of Pisces in the earliest surviving Russian star atlas, produced in 1829 by Kornelius Reissig.

SEEING STARS

Judging the number of stars has always been, and remains, tricky. Astronomer Dorrit Hoffleit at Yale University calculated that in perfect conditions – a clear night with no cloud cover, the tiniest sliver of moon and no light pollution – 9,096 stars are visible to the naked eye, approximately half visible in each hemisphere. She based the calculation on the assumption that stars of magnitude +6.5 and lower can be seen by a naked-eye observer in ideal conditions. Someone with exceptionally good eyesight well adapted to dark conditions (such as our ancestors, perhaps) might be able to see up to magnitude +8 which would increase the number of visible stars to about 40,000 (20,000 per hemisphere). With the levels of light pollution common in the industrialized world, the number of stars now commonly visible falls to around 450 per hemisphere. In highly light-polluted locations such as London or Chicago, fewer than 35 stars can be seen.

MAKING PICTURES

Although the pictures we discern in the stars are commonly referred to as 'constellations' they are more properly termed asterisms. The International Astronomical Union (IAU) uses 'constellation' more precisely to mean one of 88 specific areas of sky. These continuous divisions of the celestial sphere were formalized in 1992 by Henry Norris Russell. To a large degree they coincide with (or at least contain) the classical Graeco–Roman constellations.

The stars which we group together in asterisms might look like near neighbours, but some are far deeper in space than others. As the stars are moving away from the solar system in different directions, the constellations we now see do not quite match those seen in Ancient Mesopotamia or Ancient Greece. Over the coming millennia, the stars that make up the constellations will drift even further apart as we view them.

Pictures in the sky

The Sumerians depicted starry constellations on clay tablets from around 3200BC, and the Babylonians recorded the names of constellations (often using Sumerian names) from around 1100BC.

The Babylonians recognized around 50 constellations, several of which are familiar to us (see page 24). Early on, there were two separate traditions of rurally themed and divine constellations. The rurally themed tradition helped to provide a farming calendar for the Sumerians and later the Babylonians. The divine scheme eventually provided the signs of the zodiac and was passed on to the Greeks, so forming the basis of the Western tradition. At least some of the rural constellations might have continued in Bedouin astronomy during the 1st millennium AD.

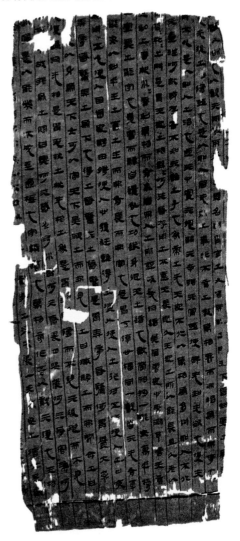

This astronomical text painted on silk was discovered at the Mawangdui tomb site in Changsha, China, in 1973. The tomb was sealed in 168BC.

Dividing up the stars

The earliest surviving star catalogues were produced in Mesopotamia, home to the Sumerians and later the Babylonians. These are the 'Three Stars Each' catalogues preserved on clay tablets from Babylonian Mesopotamia around 1100BC. According to the Babylonian creation myth, the god Marduk established order in the heavens, setting up the constellations, the division of the year into months and the allocation of different realms in the sky to different gods. He set up 'three stars each' for the months (stars that would rise heliacally, or just before dawn, in the month), giving 36 stars in total.

With only 36 stars, the 'Three Stars Each' are rather scanty catalogues, and some of those 'stars' turned out to be planets or were inaccurately located. The later MUL.APIN (see page 21) is much more extensive. It gives rising and setting dates, and names 66 stars and constellations.

When the Greeks took over the astronomical legacy of the Babylonians and Egyptians, they went far beyond the achievements of their predecessors. Around 370BC, the Ancient Greek mathematician and astronomer Eudoxus of Cnidus set out a full list of the classical constellations, describing them and the positions of the stars and including information on when they would rise and set. He is the first named author of a star catalogue. Although the original text has been lost, the poet Aratus revised it in a verse called *Phaenomena* in the 3rd century BC. Timocharis, who according to Ptolemy worked in Alexandria between 280–290BC, made the earliest Greek observations of the positions of the stars that can be dated with confidence.

When the great Greek astronomer Hipparchus produced his own star catalogue in 129BC, he found that the positions had changed slightly since Timocharis's account around 150 years earlier. This led him to the discovery of axial precession (see page 18) and of how to calculate the movement of the stars relative to the Earth. He came up with a figure of not more than one degree of axial precession per century (or 36,000 years to move through a full circle – the actual figure is about 26,000 years).

Hipparchus's star catalogue formed the basis of Ptolemy's astronomical manual *Almagest*. Ptolemy's

Ptolemy's star catalogue in the Almagest was used long after the positions he gave for the stars were no longer accurate.

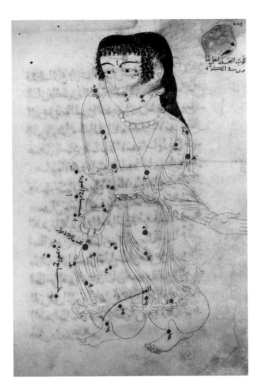

The constellation Virgo, from The Book of the Constellations and Fixed Stars *by al-Sufi (10th century).*

catalogue continued to be used in the Western and Arab worlds for around 800 years. The Persian astronomer Abd al-Rahman al-Sufi (903–983) updated it in 964, in *The Book of the Constellations of the Fixed Stars*, correcting many errors in Ptolemy's catalogue. Al-Sufi's names for some of the stars are still in use today. They were drawn from ancient Bedouin traditions of astronomy that survived alongside (and often competed with) the imported Greek knowledge. His work describes the positions, colour, brightness and magnitudes of stars, with drawings of the constellations. Most importantly, it also includes the first descriptions of Andromeda, which he called the 'little cloud', and the Large Magellanic

SET IN STONE

Although Hipparchus's star catalogue has been lost, we know that it contained the positions of at least 850 stars which he had measured using an armillary sphere (see page 73). In addition he made a celestial globe showing the constellations. Although this, too, has been lost, there is good reason to suppose that the globe carried by the Farnese Atlas – a 2nd-century Roman copy of a Greek statue – replicates Hipparchus's globe. The positions of the constellations suggest a date of 125BC ±55 years for the original, which accords with Hipparchus's dates.

HOW BRIGHT IS A STAR?

A star catalogue is not simply a list of names. Even early star catalogues often indicated the brightness of stars as well as their locations. Hipparchus and then Ptolemy classified stars by brightness, or magnitude, grouped into six categories with first-order stars being the brightest. The assumption was that the bigger the star, the brighter it is. Stars of magnitude 6 are generally considered to be least bright that can be seen with the naked eye. (In reality there is a continuum of brightness.) This method was entirely subjective, though, depending on the individual astronomer's assessment of brightness. To address this, Tycho Brahe tried to measure stars in terms of their angular size, with first-magnitude stars measuring 2 arc minutes and sixth-magnitude stars measuring $1/3$ of an arc minute. As soon as the telescope was invented, these measurements were found to be wrong, with the stars appearing much larger to the naked eye than they really are. It all took a turn for the worse as the earliest telescopes showed stars as a disc, so astronomers continued to think that the physical size of the star was visible to them. Hevelius produced a table of star sizes based on their image through these telescopes, ranging from 6 arc seconds (first-order stars) to 2 arc seconds (sixth-order stars). As telescopes improved, the artefact of the disc disappeared.

In 1856, English astronomer Norman Pogson (1829–91), working at the Madras Observatory in India, modified and quantified the Greek system. Thanks to modern telescopes, he found that first-magnitude stars are 100 times brighter than sixth-magnitude stars. This makes the difference in each step of the magnitude $100^{1/5}$, or about 2.5. This is now known as Pogson's ratio. Finally, as the distance to stars could be measured reliably by parallax (see page 107), it became clear that brightness does not correlate simply to size or distance. Today, the brightness of stars is still measured according to Pogson's ratio, but the best telescopes can detect stars down to 30th-magnitude. The assessment is made by comparing the star under investigation with a reference spot that can be adjusted until it matches the star. (There is an ultimate limit of 32nd-magnitude, imposed by the visible limit of light.)

growing. It was not long before compiling a star catalogue was a task well beyond the capability of a single astronomer and would have taken his or her entire working life.

Star catalogues became cooperative efforts but still represented a gargantuan task. The most comprehensive of the pre-photography catalogues was the *Bonner Durchmusterung* ('Bonn sampling') and its follow-up volumes, produced between 1852 and 1859 and covering 320,000 stars.

'The fixed Stars *appear to be of different Bignesses, not because they really are so, but because they are not all equally distant from us. Those that are nearest will excel in Lustre and Bigness; the more* remote Stars *will give a fainter Light, and appear smaller to the Eye.'*

Scots mathematician
John Keill, 1736

Until the mid-19th century, the only way of distinguishing between stars was by their brightness or apparent size. But after the development of spectroscopy (see page 120), cataloguing the stars took a different turn. They could now also be classified by composition.

Reclassifying the stars

The first scheme for classifying stars by spectra was devised by the Italian astronomer Angelo Secchi (1818–78). By 1866 he had developed three classes of star, numbered I–III: white and blue stars (I), yellow stars (II) and orange stars (III). (The colours of the stars are determined by their composition, with white/blue stars having the most hydrogen.) In 1868 he discovered red carbon stars, and created a new category for them (IV). He added a final class, V, in 1877, but by this time his classification was already being superseded.

Jesuit priest and astrophysicist Angelo Secchi was the first to classify stars by their composition.

Williamina Fleming led the team of female astronomers who produced the Draper star catalogue.

In 1872, the American photographer Henry Draper (1837–82) took the first photograph of the spectral lines of a star, Vega. He set out to compile a catalogue of spectra, and took 100 more photographs, but his death in 1882 curtailed the work. Two years previously, in 1880, the American astronomer and physicist Edward Pickering (1846–1919) had developed a method for photographing the spectra of many stars simultaneously by putting a glass prism in front of the photographic plate. Draper's widow donated a large sum to the observatory at Harvard College to ensure her late husband's project could go ahead, and Pickering took it over. During the coming years, Pickering's team photographed and classified 10,351 stars. Most of the classifying was done by Williamina Fleming (1857–1911). Originally from Scotland, Fleming had travelled to America with her husband and child, but her husband abandoned her. She became a maths teacher

THE HARVARD COMPUTERS

Pickering decided to employ a large team of women to do the tedious but skilled work of examining thousands of spectra from stars and carrying out the necessary calculations for the Henry Draper Catalogue. The women were officially called 'computers', but became known as 'Pickering's harem' – a patronizing term for one of the greatest gatherings of talent that astronomy has seen. Among them were women who later became famous and successful astronomers in their own right. In addition to Annie Jump Cannon (see opposite), they included Henrietta Swann Leavitt (1868–1921) and Antonia Maury (1866–1952). Cannon classified more stars than anyone else has ever done – 350,000 in total, including 300 variable stars. Leavitt's discovery of a link between a type of star called a Cepheid variable and its brightness and distance from Earth would enable scientists to calculate the distance to other galaxies too remote to be measured by parallax. Maury's work on stellar classification was a major influence on Ejnar Hertzsprung (see opposite).

and, desperate for more work, also became Pickering's housekeeper. Dissatisfied with the mathematical skills of his male employees, Pickering famously said 'My Scottish "maid" could do better' and promptly hired her and trained her in spectral analysis.

Fleming refined Secchi's system into the scheme used for the catalogue, based on the proportion of hydrogen in each star. The first catalogue was published in 1890. Pickering was so pleased with Fleming's work that he recruited more women, all highly educated. They would not normally have been able to use their skills and knowledge in the male-only world of astronomy. Fleming was put in charge of the team.

The women of the Harvard Observatory, known as 'computers', who carried out the calculations essential to the comprehensive star catalogue.

In 1901, Pickering published a catalogue of stars in the southern hemisphere with a team led by Annie Jump Cannon (1863–1941), who developed a new system of classification based on the temperatures of the stars (derived from their spectra).

Discoveries made by members of his female team prompted Pickering to begin a new catalogue. Work on the Henry Draper Catalogue began in 1911. Between 1912 and 1915, Cannon and her team classified around 5,000 stars each month. Published in nine volumes, the catalogue gives the positions, magnitude and spectral class of 225,300 stars. Pickering died in 1919, but Cannon continued the work, adding 46,850 further stars in an extension to the catalogue. She continued to record stars until her own death in 1941, when Margaret Mayall (1902–95) took over the project.

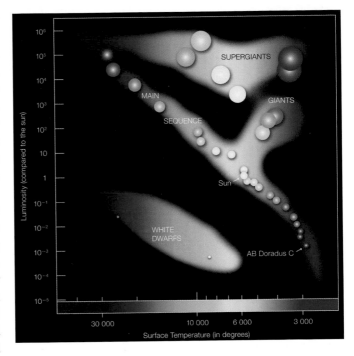

The Hertzsprung-Russel diagram maps the temperature and luminosity of stars, showing how main sequence stars change over their life.

Understanding stars

The categorization of stars by chemical composition soon led to speculation about how their composition corresponded to essential differences between stars.

The classes and composition correlate with stars of different colours and types. The Danish chemist Ejnar Hertzsprung (1873–1967) worked with the spectra catalogued in Harvard, discovering the relationship between their absolute magnitude (determined from their luminosity and a parallax calculation) and their effective temperature (from the spectra). Henry Norris Russell (he who later persuaded Cecilia Payne-Gaposchkin to drop her ideas about the composition of the Sun, see page 120) was working along the same lines. Both came up with a scatter diagram which showed stars against two axes, revealing the relationship between spectral type, luminosity and the stage in the star's development. The diagram, called the Hertzsprung–Russell Diagram, is still used in several forms as a guide to the stages of a main-sequence star's life.

DWARFS, GIANTS AND IN BETWEEN

A star is on the 'main sequence' during its period of normal helium-forging activity, from creation until it runs out of hydrogen. What happens next depends on its size. A star up to eight times the mass of the Sun loses a lot of its outer material, making a planetary nebula and a white dwarf. A white dwarf has the mass of the Sun, but the size of the Earth. A star found in 1783 by William Herschel was identified as the first white dwarf in 1910 when its anomalous size and spectra revealed that it was a previously unknown type of star.

A more massive star becomes a red giant or super-giant at the end of its main-sequence life, expanding to engulf any planets around it and growing to a massive size. It finally blows apart in a supernova, leaving a tiny, dense neutron star at its centre. The very largest produce a black hole rather than a neutron star.

Fuzzy stars

Poor eyesight or inferior telescopes can make any star look cloudy, or fuzzy, but some fuzziness is a sign of something more interesting. Ptolemy recorded five cloudy or fuzzy-looking stars, which eventually turned out to be star clusters or pairs, and a cloudy area that he did not associate with any star but which correlates with the area of Coma Berenices, a small constellation near Boötes, which contains the bright star Arcturus. These fuzzy objects became known as 'nebulous' (cloudy) objects, a name that related to their appearance rather than their nature, which was not then known.

Clouds in space

In 964, Abd al-Rahman al-Sufi (see page 155) mentioned nebulous objects in *The Book of the Constellations of the Fixed Stars*. He referred to 'a little cloud' in his description of Andromeda, which is the Andromeda Galaxy M31, and mentioned the White Ox which is the Large Magellanic Cloud. He also mentioned a nebulous star which might be the Omicron Velorum cluster IC 2391 and a 'nebulous object' now known as al-Sufi's or Brocchi's Cluster. Although he had no idea that the White Ox and the little cloud in Andromeda were galaxies, these

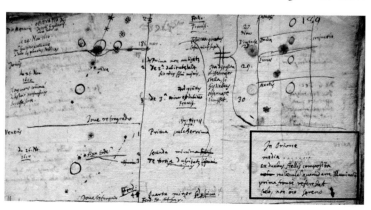

Peiresc's notes on his first observation of the Orion nebula in 1610.

remain the earliest mentions of galaxies other than the Milky Way (see page 157).

The invention of the telescope revealed more and more nebulous objects. The French astronomer Nicolas-Claude Fabri de Peiresc (1580–1637) discovered the Orion Nebula in 1610. Huygens studied it in detail in 1659, unaware that he was not the first to find it. In 1715, Edmund Halley published a list of six nebulae, and the number grew steadily until William and Caroline Herschel published a three-volume list of 2,510 nebulous objects in 1786–1802.

Clouds take shape

Some of these fuzzy objects were identified as spirals rather than just amorphous clouds, but still no one knew what they were. Then, in 1750, the English astronomer Thomas Wright (1711–86) suggested in *An original theory or new hypothesis of the Universe* that the Milky Way is a huge flat layer of planets and stars; he proposed that it looks the way it does because we are in the midst of it. This brilliant insight was taken up and popularized by the German philosopher Immanuel Kant (1724–1804), who is sometimes credited with another of Wright's suggestions – that the fuzzy, nebulous objects might be other galaxies, rather like ours, but so far away that they are not resolved into groups of stars. Wright recognized how insignificant this made the Earth:

'In this great Celestial Creation, the Catastrophy of a World, such as ours, or even the total Dissolution of a System of Worlds, may possibly be no more to the great Author of Nature, than the most

> 'Already . . . [Herschel] has discovered fifteen hundred universes! How many more he may find who can conjecture?'
> Fanny Burney, English novelist, 1786

common Accident in Life with us, and in all Probability such final and general Dooms Days may be as frequent there, as even Birth-Days or Mortality with us upon this Earth.'

The idea was quickly accepted. It must have been an astonishing paradigm shift.

From fuzzy patch to 'true' nebulae

Originally, the name 'nebula' was given to any diffuse, cloud-like object. It therefore included many objects now recognized as galaxies, or even just star clusters, as well as those which turned out to be nebulae in the modern definition – interstellar clouds of dust and gas. Herschel initially believed all nebulae to be clusters of stars too distant to be resolved into individual points, but he revised his view in 1790 when he found stars surrounded by nebulosity. These he deemed 'true' nebulae.

The dying guest

On 4 July 1054, Chinese astronomers recorded the sudden appearance of a very bright, new 'guest star'. Four times brighter than Venus, this star was visible in the daytime for 23 days and in the night sky for 653 days, slowly growing dimmer. The astronomers were actually witnessing a not-so-recent death: the aftermath of a star exploding 6,500 years previously. The 'guest star' was a supernova, now called SN 1054.

Indigenous American rock art in Chaco Canyon, New Mexico, believed to show the 1054 supernova.

SN 1054 was also noticed by Japanese and Arab astronomers, and probably by native North Americans.

Back from the dark

After its brief 653 days of glory, SN 1054 disappeared from view. No trace of it was seen for nearly 700 years. Then, in 1731, with the benefit of a telescope, the British astronomer John Bevis discovered the detritus from SN 1054. In 1758, French astronomer Charles Messier (1730–1817) was looking for Halley's comet when he found a blurry smudge the shape of a candle flame. At first he thought he'd found the comet, but instead he had stumbled on the remnants of SN 1054.

As Messier's intention was to find comets, fuzzy objects such as SN 1054 were a nuisance. He decided to make a catalogue of them to help other comet-hunters. His final edition recorded 103 objects including supernovae, diffuse nebulae, planetary nebulae, open clusters, globular clusters and galaxies. More objects have been discovered since, giving the total of 110 Messier objects now recognized. The Crab Nebula is M1. It was named the Crab Nebula by the Anglo-Irish astronomer

SUPERNOVA ➤ NEBULA + PULSAR

After a massive star has used up most of the hydrogen that fuels it, it is no longer dense enough to hold together. When it reaches a critical point, the central part collapses in a huge explosion. It's all over in seconds. Most of the star's mass is blasted away into space, producing the brilliant explosion of light witnessed as a supernova. What is left is a very dense neutron star only 28–30km (17–18 miles) across in the case of SN 1054. The remnants of the old star drifting away form a cloud of debris we see as a nebula.

When the star collapses, it goes from normal to very rapid rotation (up to 42,000 times a minute). The energy and magnetic field of the neutron star reach Earth in rapid pulses as the star rotates – hence the name pulsar. The neutron star left by SN 1054 was one of the first pulsars discovered in 1968 (see page 169). It's often used to calibrate flux density in X-ray astronomy (that is, calibrate the flow of X-ray energy through a particular area) giving units of 'crab' and 'millicrab', from its name, the Crab Nebula.

The Crab Nebula, the remains of the star that exploded in 1054, is made up of dust, hydrogen and helium forming a cloud 11 light years across and is still growing at a rate of about 1,500km (932 miles) per second.

resembles a planet which is fading'. For the comet-hunters, though, these objects were just getting in the way.

Herschel picked up on the resemblance to a planet in 1782, describing a nebula as a 'very bright, nearly round planetary disc.' The mistaken association with planets has left us with the name 'planetary nebula' for the expanding shell of gas encircling a dying

William Parsons (Earl of Rosse) in 1844, who noticed that his drawing of it looked like a crab, but with extra stringy bits. Rosse (1800–67) built the largest telescope in the world, the 1.07m (42in) 'Leviathan of Parsonstown', so was able to produce detailed images.

It might seem odd that Messier and his colleagues did not want to investigate the objects they had found. Antoine Darquier described one found in 1779 as 'very dull, but perfectly outlined; it is as large as Jupiter and

The catalogue of Messier objects now contains 110 objects; Messier himself found only 17.

star. The object has nothing to do with planets, but the name has stuck.

Well-timed guests

The only two comparable supernova events since 1054 were astonishingly well timed, occuring in 1572 and 1604, just as the revolution in astronomy was taking off. If the 1604 event had occurred five years later, Galileo would have been able to witness it with his telescope. The supernova of 1572 was witnessed by Tycho Brahe and others in Europe. In Ming dynasty China, it was considered an ill omen, perhaps a warning to the young emperor to mend his ways.

Tycho's supernova (SN 1572) was incontrovertible evidence that the heavens are not eternal and immutable. Here was a star appearing from nowhere, and disappearing again later – and not just a comet, which moves across the sky and was deemed to be lower than the stars and planets, but a star that belonged in the same sphere as the others. That another should come so soon after, just 32 years later, was an astonishing piece of good fortune for the history of astronomy.

Lines and fuzz

It was not possible to tell how the various types of Messier object differed until the development of spectroscopy (see page 120). Then in 1864 the English astronomers William and Margaret Huggins examined the spectral emission and absorption lines of several nebulae. The potential of spectral lines – Fraunhofer lines as they were known – was first explored by amateur astronomers, many of them living in and around Victorian London. One of the most important was Margaret Murray, who had learned some astronomy from her grandfather and built her own

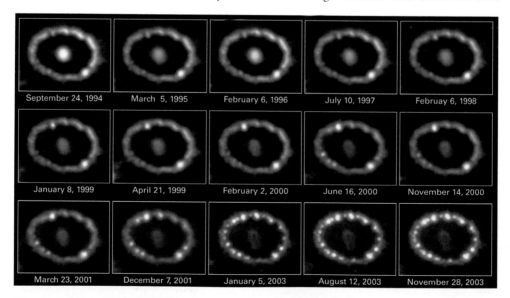

September 24, 1994	March 5, 1995	February 6, 1996	July 10, 1997	Februay 6, 1998
January 8, 1999	April 21, 1999	February 2, 2000	June 16, 2000	November 14, 2000
March 23, 2001	December 7, 2001	January 5, 2003	August 12, 2003	November 28, 2003

These Hubble Space Telescope images of Supernova 1987A shows changes in the exploded star between 1994 and 2003.

EXPLODING STARS THROUGH HISTORY

There have been only eight supernovae visible to the naked eye within the Milky Way in recorded history. The first was in AD185. Chinese astronomers described it as visible for several months and looking like a bamboo mat displaying five colours, 'both pleasing and otherwise'. Subsequent supernovae were recorded in AD 393, 1006, 1054, 1181, 1572 and 1604. One in the mid-17th century might have been recorded without being identified clearly at the time; there was a reported 'noon-day star' in 1630, and John Flamsteed catalogued a star in the right position in 1680 to match a remnant, Cassiopeia A, discovered in 1948. That of 1006 was possibly the brightest recorded star ever to be seen in the night sky, reported in Egypt by Ali ibn Ridwan (988–1061) as being a quarter the brightness of the Moon. The supernova of 1181 was no brighter than a first-magnitude star, but was recorded by Chinese and Japanese astronomers. Chinese records refer to further events that might have been supernovae but have not so far been linked with any remnants.

There have been no supernovae visible to the naked eye since the 17th century, but that does not mean they are not happening. Astronomers estimate that a supernova occurs about every 50 years in the Milky Way.

spectroscope. In 1875, she married another amateur astronomer, William Huggins. He had been a successful young businessman, but sold his business and bought a good telescope so that they could follow their passion for spectroscopy. They published their *Atlas of Representative Stellar Spectra* in 1899.

The couple noted that while about a third of nebulous objects had the emission spectra characteristic of clouds of gas, the rest had spectra characteristic of stars. This was the first evidence that some are groups of stars, while others are vast clouds of dust and gas (true nebulae). But it was far from compelling proof of the nature of the starry nebulae, and the discussion became heated over coming years.

Making stars

It is now known that nebulae can be the cradle of stars as well as their graveyard.

Towering columns of interstellar gas and dust collapse as gravity pulls them together, forming first nebulae and then stars. These 'star nurseries' were only properly observed with the advent of space-based telescopes such as the Hubble Space Telescope. Before that, though, the role of supernovae and vast gas clouds in recycling the material of stars became clear.

Hans Bethe had explained how stars produce helium from hydrogen, but what about everything else? The spectral lines from the Sun and other stars also showed the presence of much heavier elements. That puzzle was solved by the British astronomer Fred Hoyle in 1946. His theory of nucleosynthesis proposed that all elements are forged within stars. Essentially, at the end of a star's life, when it runs out of hydrogen, it begins to force helium nuclei together. Three helium nuclei come together, forming carbon. Even later, as the star runs out of helium, what happens depends on its size. In the largest stars, the carbon is forced together into heavier elements, and this process continues until the core is iron. Finally, that collapses producing a neutron star or a black hole (see below), with everything else, the great mix of elements, blasted off into space. The energy of the blast itself makes even more new elements. And from the star debris, much later, planets with the composition we see in the solar system are formed. We really are stardust.

Starting from Fraunhofer's spectral lines, the composition of the stars and the existence of all the elements in the universe were explained. Hoyle, an atheist, had managed to explain the creation of stars and worlds without recourse to any supernatural cause. Thales would have approved.

The death of stars

As we have seen, a large star can end in a supernova and pulsar. An even larger star can result in a black hole. The star collapses in on itself to the extent that all its matter is squeezed into a very tiny space and its gravity is so extreme that not even light can escape from it. The event horizon of a black hole is a boundary beyond which nothing can escape, but is drawn in by the black hole's gravity. The theoretical possibility of black holes emerged from Einstein's relativity equations, as various astronomers worked on them from the 1920s onwards. But Pierre-Simon Laplace had proposed something similar in 1796, postulating a star so massive, with so much

The 'pillars of creation' are towering columns of gas and dust 6,500–7,000 light years away. New stars are formed from this mix of materials.

gravity, that not even light could escape from its surface.

At the centre of a black hole is a 'singularity', a point where the curvature of space-time becomes infinite. In 1939, American physicist Robert Oppenheimer predicted that neutron stars with mass greater than three times the mass of the Sun would eventually become black holes. Research into black holes continued in the 1950s and 1960s, but was entirely theoretical – based on mathematics, with no evidence of any black holes or neutron stars existing.

Then, in 1967, the Northern Irish radio astronomer and astrophysicist Jocelyn Bell Burnell (b.1943) found a pulsar. While working on quasars, radio-bright objects found at the heart of galaxies, she came across a rapidly pulsing radio signal she could not identify. She labelled it LGM-1 (for 'little green men', in case it was a signal from aliens). Investigation eventually revealed that it was a neutron star rotating at approximately one revolution per second: in other words, a pulsar.

Theoretical evidence for black holes became more compelling with the work of English astrophysicists Roger Penrose (b.1931) and Stephen Hawking (b.1942) in the 1960s and 1970s. In 1970, they published work on the existence of singularities in black holes and stated that if the Big Bang theory of the origins of the universe is correct (see page 184), then the universe originated with a singularity. In 1974, Hawking demonstrated that black holes can emit radiation, now called Hawking radiation, and through this process can eventually exhaust themselves and evaporate.

Galaxies disputed

From the middle of the 19th century, as telescopes improved, more details of the nebulae became visible. In 1845, William Parsons noticed that Messier 51 has a spiral shape. (It is now known as the Whirlpool Galaxy.) It was the first spiral nebula to be discovered, but others soon followed.

Some astronomers suggested that as there are quite a few spiral nebulae, it might be the case that the Milky Way is also one of them. If the Milky Way were a relatively flat spiral, this would explain its appearance as a band across the sky as we would be in the plane of the band. This was a development of Thomas Wright's suggestion in 1750 that the Milky Way is a layer of stars and we are in its midst. Other astronomers maintained that the spiral nebulae being discovered were simply great clouds of gas and dust, not other galaxies at all. There was no agreement, and insufficient evidence to decide. It might not sound like a hugely important question about the nature of some spiral-shaped cloudy objects, but it had significant implications. If the spiral nebulae were other galaxies, then the size of the universe was far, far greater than anyone had previously imagined.

The Great Debate

By the early 20th century, opinion was divided between those who believed spiral nebulae were vast clouds of gas within the Milky Way and those who believed them to be separate 'island universes' – galaxies – outside it.

In 1918, the American astronomer Harlow Shapley (1885–1972) used a technique based on calculating distances

If we could step outside the Milky Way, it would probably look very like the Andromeda galaxy.

to several stars of known luminosity to work out the position of the Sun within our Milky Way galaxy. He found that we are about 30,000 light years from its centre. He believed the Milky Way to be so vast that nothing could lie outside it.

Another American astronomer, Heber Curtis (1872–1942), believed that spiral nebulae are other galaxies and lie outside the Milky Way. In 1920, a debate between Shapley and Curtis set out both sides of the argument. Known as the Great Debate, it was a landmark in 20th-century astronomy. It took place at the Smithsonian Museum of Natural History in Washington, DC, USA.

Both men were right about some things and wrong about others. Shapley was right about the Sun being off centre in the Milky Way (Curtis believed it was central) but wrong about spiral nebulae being gas clouds within our galaxy. Curtis was right about spiral nebulae being other galaxies. The question was finally resolved by Edwin Hubble (1889–1953). Using the 254cm (100in) Hooker telescope on Mt. Wilson, California, he produced photographs with deep exposures on large fields of stars and resolved individual stars within the Andromeda galaxy. Following the same technique as Shapley to calculate distances, Hubble found that these stars were more than ten times further from the Sun than even the most distant stars of the Milky Way, so must be in another galaxy. He estimated the distance from the Sun at a million light years. (Andromeda is actually around 1.5 million light years away.) Andromeda became the first other galaxy to be positively identified. Hubble published his results in 1929, changing forever our view of the universe and our place in it. In the space of around 400 years, discoveries in astronomy and biology had demoted humans from supreme, created rulers at the centre of the universe to evolved and evolving beings on one small planet orbiting an off-centre star in one of many galaxies.

GALAXIES GALORE

Since Hubble's discovery, astronomers have found many more galaxies. In 2016, a whole host of previously unknown galaxies was found hiding behind the Milky Way. It had been difficult to 'see' behind the bright distraction of the stars in our own galaxy, but the radio telescope at the Parkes Observatory in Australia revealed 883 galaxies lurking in the so-called Zone of Avoidance. There are thought to be 100–200 billion galaxies in total. This is calculated by using the Hubble Space Telescope to examine a tiny region of space over hundreds of hours, so allowing even the faintest light from another galaxy to be collected. Astronomers then extrapolate from the number of galaxies observed to get a value for the whole sky. A galaxy such as the Milky Way has 200–400 billion stars; many galaxies are smaller and have fewer stars, but some are larger and have trillions of stars. That suggests perhaps 4×10^{22} stars altogether. Again, that's just the observable universe; the whole universe could be much larger, or even infinite (see page 189).

CHAPTER 7

Remaking the
UNIVERSE

'It is established by evidence that there exists
beyond the world a void without a terminal limit,
and it is established as well by evidence that God
Most High has power over all contingent beings.
Therefore He the Most High has the power to
create a thousand thousand worlds beyond this
world such that each one of those worlds be bigger
and more massive than this world.'

Fakhr al-Din al-Razi,
Matalib al-'Aliya, 12th century

**By the end of the 17th century, the
Copernican model had provided an
entirely new framework within which
the mechanics of the universe were
worked out. But telescopes led to the
extent of the known universe growing
ever larger and revealing an increasing
number of phenomena which demanded
explanation. This required a new
approach to cosmology, one rooted in
mathematics.**

*First mentioned by al-Sufi in the 10th century, the Large
Magellanic Cloud was one of the first galaxies outside the Milky
Way ever described.*

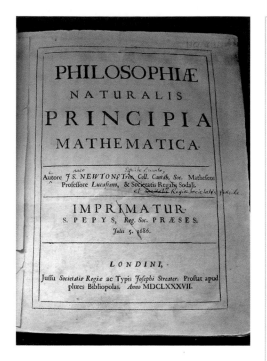

Newton's Principia *was finally published in 1687, 20 years after he formulated his ideas about gravity.*

distance between them, so if the distance doubles, the force reduces to a quarter. He explained why the path of bodies in orbit is elliptical, and he applied this to planets, moons and comets. Newton did not require any concept of *aether*, as gravity acts at a distance with no need to be mediated by any type of matter in between. His theories drew together phenomena as diverse as the tides, the orbit of comets, the path of the Moon and axial precession. It provided an explanation for the motion of all heavenly bodies, within and beyond the solar system.

It took a little time before Newton's account was universally accepted. Many scientists preferred Descartes' system. It had the advantage of simplicity – it needed God to set things in motion and the rest just followed the laws of physics without the need for the mysterious force of gravity working at a distance. Furthermore, it didn't contravene scriptural teachings that the Earth is stationary. As Descartes described it,

> '*All matter attracts all other matter with a force proportional to the product of their masses and inversely proportional to the square of the distance between them.*'
>
> Isaac Newton, 1687

ISAAC NEWTON (1642–1727)

Isaac Newton had an unhappy, disrupted childhood and was described as 'idle' and 'inattentive' at school. He went to Cambridge University to study law, but became interested in mechanics, physics and astronomy, teaching himself mathematics. In 1665, plague closed the university and Newton returned home to Lincolnshire. It was a productive time for him. He developed the mathematical method of differential and integral calculus, discovered that white light can be split into a spectrum of colours and began his work on gravity and the laws of mechanics.

Back in Cambridge, Newton was appointed Lucasian professor of mathematics in 1669 at the age of 27 (a position later held by the astrophysicist Stephen Hawking). He taught that light is made of particles ('corpuscles') rather than waves; this had long been a point of dispute among scientists. His most important work was on physics and the movement of astronomical bodies.

Principia made Newton internationally famous and within 50 years his theory was universally accepted. Newton's account of gravity prevailed until the 20th century when it was superseded by the work of Albert Einstein (see page 179). Alongside his work in science, Newton keenly pursued his interests in alchemy, theology and ancient history.

Newton was a difficult man. He disliked engaging with other people, could not deal with professional disagreement and tended to be argumentative. He developed a life-long animosity towards Robert Hooke, and later Gottfried Leibniz. He was reluctant (and slow) to publish his work, apparently in order to avoid confrontation. Newton became president of the Royal Society in 1703 and remained president until his death. In 1705, he became the first scientist ever to be knighted for his work.

'The Earth, properly speaking, is not moved, nor are any of the Planets; although they are carried along by the heaven.' Newton also claimed to have God working his model, though he could not lay claim to a stationary Earth. In Newton's model, God created the universe to follow physical laws, set it in motion, and then didn't need to do anything – but gravity was still God's invention. This 'clockwork-universe' model fitted well with the mood of the Enlightenment, and by the mid-18th century, the Newtonian gravitational model was generally accepted by most people of science.

Refining Newton

A particular problem concerned astronomers at the time: Jupiter's orbit appeared to be getting smaller while Saturn's appeared to be growing. The French mathematician and astronomer Pierre-Simon Laplace (1749–1827) developed the mathematics he needed to solve the problem. His solution was based on the fact that two of Saturn's orbits are nearly equal to five of Jupiter's, meaning that the planets have a close approach about once every 900 years. This is enough to create perturbations in the orbit. With Laplace's calculations, the tables predicting the orbits of the two became much more accurate.

Laplace attempted to extend his model for Jupiter and Saturn to the entire solar system, with only limited success. Indeed, the problem of how multiple orbiting bodies influence one another is not solvable. His summary work on the solar system, *Exposition du système du monde* and the *Mécanique celeste* ('Explanation of the world system' and 'Celestial mechanics') presents all the calculus behind Newton's model, filling in details Newton could not complete.

More movement

While Newton and those who came after him focused primarily on the mechanics of the solar system, William Herschel took a broader view. He was the first to measure the proper motion of stars, their movement relative to one another (see box, page 180), and to realize that they were moving closer

Pierre-Simon Laplace was the first to tackle the problems of celestial mechanics using calculus.

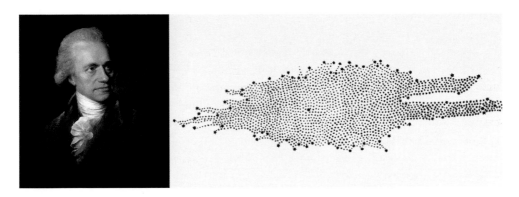

William Herschel (above left) produced a map of the Milky Way by recording the positions of the stars.

together in one region but further apart in another. Proper motion had been suspected much earlier. Macrobius recorded around AD400 that some Greek astronomers believed the size of the universe and distance of the stars are the only reasons why we cannot see the stars moving. They do move, but from where we are standing their movement is so gradual that we can't detect it.

In 1718, Edmund Halley first noted the proper motion of three bright stars: Sirius, Aldebaran and Arcturus. Comparing his own measurements of their positions with those recorded by Hipparchus around 150BC, he found a noticeable change. Some stars had moved further than others. For example, Sirius had progressed half a degree (about the diameter of the Moon) to the south in 1,850 years.

Later that century, Herschel also became interested in the proper motion of the stars, publishing his findings in 1783. He concluded that the Sun (and therefore the Earth in orbit around it) is moving towards those stars which seem to be moving apart – because as we get closer to them, distances between them look larger. This was a striking finding as it showed that the Sun is not at the still centre of the universe after all; Herschel found it to be moving on its own trajectory, towards the star Lambda Herculis.

The bigger picture

Even as the universe expanded with more and more stars, nebulae, and possibly even new galaxies, Newton's basic account of planetary mechanics held good, augmented by the refinements of later astronomers such as Laplace. Then in 1905, an Austrian patent inspector threw a spanner in the celestial works.

Relativity replaces Newtonian mechanics

Albert Einstein (1879–1955) was not an astronomer but a theoretical physicist, yet his work has had a profound effect on cosmology. His theory of general relativity has effectively replaced Newton's explanation of gravity. In one remarkable year, 1905, he published four ground-breaking papers, including his theory of special relativity. His most important findings for astronomy were that:

PROPER MOTION OF THE STARS

Proper motion is the movement of a star in relation to others over an extended period. (It is measured in arc seconds per year.) 'Proper', in this sense, means the stars' 'own' motion. Unlike the apparent movement caused by the Earth's axial precession, proper motion is cumulative, increasing over time as the stars genuinely move in relation to the Sun. How much a star appears to move depends on its direction of movement with regard to the Sun (and therefore to us). If a star is moving directly away from or towards us, no proper motion is apparent. If it is moving in any other direction instead or as well, it will show greater or lesser motion depending on the plane of its movement and how far away it is. Closer stars tend to have larger proper movements than more distant stars over the same period of time.

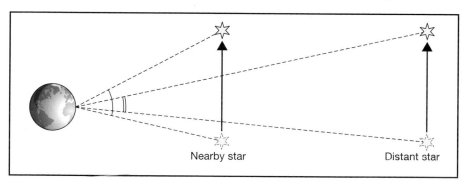

Nearby star   Distant star

- electromagnetic radiation comes in discrete packages ('quanta') of energy, rather than waves. Newton's finding that light is delivered in 'corpuscles' had since been overturned by James Clerk Maxwell's model of electromagnetic radiation as waves of energy.
- the laws of physics are the same everywhere in the universe.
- the speed of light is both a fixed constant and a universal speed limit – it is the same no matter what the observer's frame of reference, and nothing can exceed it.
- energy and matter are interchangeable, defined by the equation $E = mc^2$ where

E = energy, m = mass and c = a constant (the speed of light). Tiny amounts of mass can be converted into huge amounts of energy – a finding that later explained how stars are powered and opened the door to nuclear weapons and nuclear power.

Einstein was troubled by aspects of special relativity and continued to work on the theory, publishing his theory of general relativity in 1916. This proposed that space-time is distorted by objects with mass, producing the effect of gravity. This distortion causes one object to move towards another, with the more massive object having the greater effect. Another aspect is that gravity and acceleration are indistinguishable. The following year, he published his ideas about the 'cosmological constant' – a sort of anti-gravity force that prevented the universe collapsing in on itself as it otherwise would under the force of gravity. He later called this a 'blunder', but his theory has since been re-examined as a possible explanation for the phenomenon of 'dark energy' (see page 189).

Einstein's explanation replaced or perhaps extended Newton's theory of gravity. In Newtonian mechanics, mass, time and distance are taken as fixed. This works for most frames of reference, and

Nuclear weapons exploit the power locked within atoms and explained by Einstein's equation E = mc².

Newton's laws are good enough to predict most planetary movement and for launching spacecraft. But in Einstein's universe these are relative: they vary with frame of reference. Einstein's model applies at all size levels and in all frames of reference, while Newton's doesn't work well at very high speeds or small sizes.

Bendy light

Einstein argued, importantly, that light can be bent by gravity. The degree of bending predicted by Newtonian mechanics is about half that predicted by relativity theory.

The validity of Einstein's theory was demonstrated by a famous experiment that exploited just this difference. The English astronomer Arthur Eddington (1882–1944) took a ship to Principe off the coast of Africa to photograph a full solar eclipse in 1919. Measurements drawn from his photographs demonstrated that light from stars hidden behind the Sun is bent on its path to Earth by the Sun's gravity, so making the stars visible. The effect – light bending around a massive body – is now used by astronomers in many ways, including the search for exoplanets (planets orbiting stars other than the Sun; see page 201).

Beginning and being

Einstein's work raises again a question that had troubled even the earliest astronomers: is the universe fixed and stable, or is it changing? This question about its immediate state leads to larger issues about its past and present: is the universe

This composite image made from photos taken by the Hubble space telescope during 841 orbits of Earth shows 10,000 distant galaxies. Blue galaxies are the newest, and stars are probably being formed in them.

finite in time, with a beginning and an end, or is it eternal? If it is finite, is there just one universe, appearing only once, or is it cyclical, appearing again and again?

In the beginning . . .

Every culture has its origin myths. They are the stuff of religion and storytelling, mythic cosmologies that attempt to answer the philosophical questions of why we are here and how we came about. But for the last 2,500 years, European culture has also had theories, rather than just myths, about origins. The Ancient Greeks made the first attempt to explain the origins of the universe without recourse to the divine, offering the first proto-scientific explanation (although

they also had their own separate creation myths). Anaximander claimed that the 'boundless' has always existed and will always exist: 'all the heavens and the worlds within them' come from 'some boundless nature'. A tiny portion or 'germ' became separated from the boundless and formed a ball of fire encircling the vapours that surround the Earth. The ball of fire divided into several rings, which became the Sun, Moon and stars.

Anaxagoras proposed that the universe was a mixture of diverse ingredients divided into infinitely small fragments. The mix was set into a whirling motion by the action of nous (mind) and the movement separated out the ingredients so that they clumped into the different types of matter and objects we see around us.

Parmenides argued in the early 5th century BC that nothing can come into existence from nothing, nor disappear into nothing: the universe is therefore eternal and unchanging. Where there is not something else there is *eon*, which is the essence of being. Aristotle, in the 4th century BC, agreed that the universe is unchanging, but he did not consider it infinite in extent.

The Greek Stoics, in the 3rd century BC, adopted a cyclical model, in which the parts of the universe that we can see came into being and will eventually cease to exist and re-form in another configuration. Stoic cosmology taught that in its original state, the universe was entirely *pneuma* (breath), which was also

God. A mix of tension and heat in the *pneuma* led to the different elements distilling out – first fire, then air, then the heavier water and finally earth. From these, the matter we see around us formed. The innate tension of the universe continues, though, and will eventually destroy everything. The Earth and universe will decay, matter returning first to its elements and then to the *pneuma*. At some point, the cycle will begin again.

Despite these various models – and there were more – the eternal universe became the dominant model in the West. Like other aspects of Aristotle's cosmology, this notion appealed to the Christian Church and to Islam as it allowed God to have initiated an eternal, perfect, unchanging creation in keeping with the teachings of the holy texts. For a long time, the universe was not deemed to have beginning or end.

Stasis challenged

The unchangingness of the universe went largely unquestioned in the West until Tycho Brahe's supernova appeared in 1572. At that point the English astronomer Thomas Digges (1546–95) attempted to measure the distance to the 'new star' using parallax (see page 107). He was unsuccessful: it was too far away. As it was possible to use naked-eye parallax to measure the distance to relatively close objects, the failure of the method demonstrated that the star was beyond the orbit of the Moon, in the zone where things were not supposed to change.

The other aspect of the religious narrative, that the universe came about through an act of divine creation, endured even longer. Only after the literal interpretation of the Bible narrative had taken a severe battering at the hands of 19th-century scientists did the origin of the universe become subject to scrutiny.

So when the question of the origins of the universe came up again in the 20th century, it was around 2,000 years since many of the arguments had been properly aired. As we have seen, Einstein favoured a static universe, as Newton had done: infinite, but stable, neither growing nor shrinking. But Einstein was, it seems, wrong.

All from nothing

The Belgian priest and astronomer Georges Lemaître (1894–1966) discovered a case for an expanding universe while working

The supernova of 1572, and Tycho Brahe's explanation of it, proved that the heavens are not unchanging.

Georges Lemaître saw no challenge to his religious beliefs in the theory of the Big Bang.

that if everything is moving apart it is reasonable to suppose that it had originally all been together. He proposed that the entire universe had expanded from a single point, a 'primaeval atom' or 'cosmic egg'.

It took a few years for the cosmic egg theory to gain traction; it was unpopular to start with and ridiculed in some quarters. Indeed, the name by which the theory is now known – the Big Bang – is coined from a sarcastic remark made by the English astronomer Fred Hoyle in 1949. But eventually Einstein and others came to accept it, and it is now the most widely accepted scientific paradigm for the origin of the universe, approved even by the Church.

Pigeons or the universe?

In rejecting the Big Bang idea, Hoyle challenged supporters of the theory to find evidence of the heat – the 'fossil'

with Einstein's equations for relativity. He also proposed that the most distant galaxies would be travelling away from the solar system the fastest. He published his findings in 1927, in French, but they went unnoticed until translated into English.

At the same time, the American astronomer Edwin Hubble was examining distant galaxies at the Mt. Wilson Observatory. His findings in 1929 confirmed Lemaître's prediction: not only are galaxies moving away from Earth, but the most remote galaxies are moving away most quickly. Two years later, in 1931, Eddington came across Lemaître's paper and arranged to have it translated into English. The same year, Lemaître said at a meeting in London

---

3,000-YEAR-OLD COSMIC EGGS

The idea of the entire universe expanding from an infinitely small point is first found in the *Rigveda*, a collection of Hindu hymns written in Sanskrit in India around 1500–1200BC. The text sets out the theory (or myth) of a 'cosmic egg', Brahmanda, which holds the entire content of the universe in a single infinitely small point called a Bindu. The universe expands from the 'egg' and after a long period of time collapses into it again, before expanding again in an infinite cycle of expansions and contractions. An infinite series of expansions and collapses remains a possible model under Big Bang theory.

> 'The Big Bang, which is today posited as the origin of the world, does not contradict the divine act of creation; rather, it requires it.'
>
> Pope Francis, 2014

of the Big Bang – that should be left over from that initial explosion. This evidence was not long in coming. In 1964, American radio engineers Arno Penzias and Robert Wilson were working on a super-sensitive new radio antenna, trying to remove all sources of interference and background noise. When they had eliminated all extraneous signals, there was still faint background noise, 100 times the intensity they expected. Blaming pigeons nesting near the antenna, they removed the pigeons and their droppings. The noise persisted, and even spread across the sky, day and night. Hearing from Bernard F. Burke, a professor at Massachusetts Institute of Technology,

about work suggesting that radiation from the Big Bang might be detectable in space, Penzias and Wilson realized that they had discovered just that – Cosmic Microwave Background Radiation (CMBR), the echo of the Big Bang.

After the Bang

In 1980, the American physicist Alan Guth proposed an inflationary universe. This includes a period just after the Big Bang in which the universe expanded exponentially for a very short period of time. It accounts for some problems in the maths of the Big Bang model, and is now generally accepted. The history of the universe can now be modelled with some confidence back to the first 10^{-44} of a second (that's 0.00000000000 0000000000000000000000000000001 of a second!)

The size of the universe

The concept of a universe infinite in space is difficult to grasp even now. The question

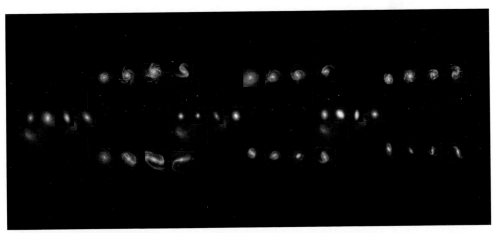

The types of galaxies that astronomers believe could be found in the universe today (left), 4 billion years ago (centre), and 11 billion years ago (right).

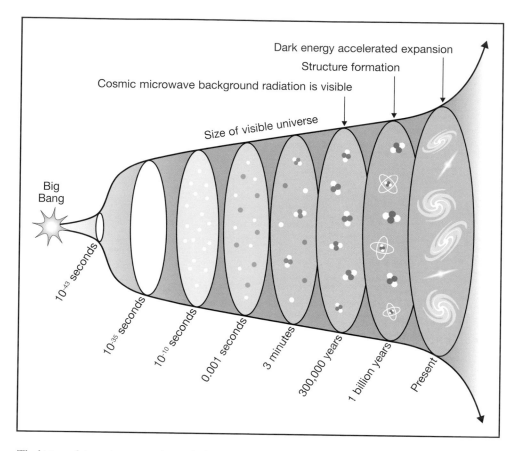

Dark energy accelerated expansion

Structure formation

Cosmic microwave background radiation is visible

Size of visible universe

Big
Bang

10^{-43} seconds

10^{-35} seconds

10^{-10} seconds

0.001 seconds

3 minutes

300,000 years

1 billion years

Present

The history of the universe, starting with the Big Bang and showing an initial phase of rapid inflation, then stabilizing and finally accelerating expansion.

'But what's outside it?' naturally springs to mind and is rarely satisfied with the answer that there is no outside.

A bounded universe

Both the Ptolemaic and Copernican models had the entire system enclosed in a sphere; the universe was bounded, finite. The fixed stars were equidistant from Earth, attached to the outermost orbiting sphere. When we look at the night sky, there is nothing to suggest that the stars are at different depths in our field of view. They are of varying brightness, certainly, but this could be because they are different in size or intensity rather than distance.

Breaking outside

Possibly the first scientist to challenge the bounded universe idea of the stars was Thomas Digges. He shattered the outermost sphere of the Copernican universe by suggesting that the stars are not in a fixed band but extend outwards into infinite space.

ARCHIMEDES RECKONS THE SIZE OF THE UNIVERSE

In a work called *The Sand Reckoner*, Archimedes set out to calculate the number of grains of sand it would take to fill the universe. He began with Aristarchus's Sun-centred universe. He had to invent a way of expressing large numbers, since the largest named number at the time was the myriad (10,000). Archimedes came up with a number equivalent to 10^{63} grains of sand. By astonishing coincidence, this is the same as the current estimate of the size of the universe, which is 10^{80} nucleons; 10^{63} grains of sand contain around 10^{80} nucleons.

How far to a star?

Whether or not many people took on board the notion of infinity, the size of space was certainly an issue. As soon as the stars were released from their fixed sphere to spread out through three dimensions, the question of the distance to and between them arose. When Galileo discovered that the Milky Way is made up of stars, the universe suddenly became much larger than previously thought – it had to be many times, perhaps thousands of times, larger to accommodate all the new stars found.

The first attempt to measure the distance to a star was made in the late 17th century by Christiaan Huygens. His method was ingenious.

Huygens drilled holes of different sizes in a brass plate, then held it up to the Sun. He hoped to find through one of the holes that the brightness of the Sun would match the brightness of Sirius, the brightest star in the night sky. (There is obviously a large margin of error in this method, as he had to compare it with his memory of the brightness of Sirius, since Sirius can only be seen at night when the Sun is not there for

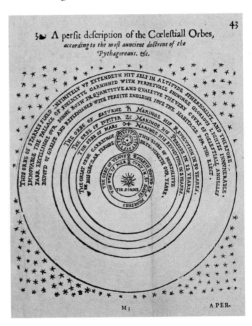

Thomas Digges' depiction of the Copernican universe with unbounded stars extending beyond the solar system.

'This orb of stars fixed infinitely up extends itself in altitude spherically, and therefore immovable the palace of felicity garnished with perpetual shining glorious lights innumerable, far excelling over [the] sun both in quantity and quality.'

Thomas Digges, 1576

comparison.) But the Sun never did match the brilliance of Sirius, no matter how small he made the holes: the Sun was always much brighter. Next, he bought opaque beads to put in front of the smallest hole, but still the Sun was too bright. Finally, when he had reduced the light of the Sun by a factor of around 800 million, he judged it to be of equivalent brightness to Sirius. From this figure, using the knowledge that brightness is inversely proportional to distance, he calculated Sirius as 28,000 times the distance from Earth to the Sun. This works out at about 0.4 light years, whereas in fact Sirius is 8.7 light years away. But Huygens' calculation wasn't far wrong, he had just made a false assumption: he had assumed that Sirius would be the same brightness as the Sun, but it is about 25 times as bright. (The same assumption, that magnitude correlates directly with proximity, caused problems for those cataloguing the stars.)

The first accurate measurement of the distance to a star came in 1838 with the work of the German astronomer Friedrich Bessel (1784–1846). Bessel recognized that the larger the parallax effect of a star, the closer the star must be. On the basis of this assumption, he chose the star 61 Cygni for his calculation. It was quite a brave choice, as it's not a bright star. He was the first astronomer to measure parallax for a star successfully, having made detailed, accurate measurements of the proper movements of 50,000 stars (see page 107).

Growing and growing

Although the universe is expanding, gravity pulls matter together. This would suggest that the expansion of the universe is limited. Einstein had already proposed a mechanism (he called it the cosmological constant) to prevent the universe collapsing under the effects of gravity. Clearly, following on from Hubble's observations (see pages 170–71), the expansion had enough energy to keep going for a while. Then in 1998 the Hubble Space Telescope, named in honour of the great astronomer, returned a surprising result. Far from slowing down, the expansion of the universe is speeding up. Two teams of astronomers who were investigating light from a distant

By taking measurements from Earth six months apart, the movement of the planet can be exploited to measure parallax.

PARALLAX AND PARSECS

To measure the parallax of a star, it is necessary to take sightings from widely separated points. Sightings can be taken from different positions on Earth, or from the same position but at different times of year (so it is the Earth itself that has changed position).

Astronomers measure the distance to stars in parsecs, which is 1/parallax angle in arc seconds. As there are 60 arc seconds in an arc minute and 60 arc minutes in a degree, an arc second is 1/3600th of a degree. The angle of parallax for stars is very small; even the very closest star, Proxima Centauri, has a parallax angle of only 0.772 arc seconds. The distance to Proxima Centauri is therefore 1/0.772 = 1.30 parsecs. One parsec is about 3.26 light years.

supernova discovered it was further away than it should be if the expansion of the universe were slowing down.

There are several explanations for this finding, one being that Einstein's theories are wrong. But the currently favoured theory is that space is being pushed apart by a mysterious force called 'dark energy'. Einstein conjectured that empty space is not nothing; it could possess its own energy. Also, more space can just come into existence, bringing dark energy with it; the amount of dark energy will keep increasing, pushing up the pace of expansion ever further.

AS FAR AS WE CAN SEE – AND FURTHER

Current estimates suggest that the observable universe is a sphere 93 billion light years (28.5 gigaparsecs) across. The sphere is centred on Earth, as we can see out in all directions equally. But that doesn't mean the entire universe is limited to 93 billion years across, and certainly doesn't mean that we are in the centre of the entire universe. Nor is the size of the universe stable, as it is still expanding.

What we know is tiny

No one knows exactly what dark energy is, but there's a lot of it. Figures released by NASA at the start of the 21st century suggest that dark energy makes up about 68 per cent of the universe. A further 27 per cent or so is dark matter – matter we can't see – which might not be 'normal' matter. That leaves only about 5 per cent for the matter and energy we do know about. There's a lot we don't know!

The end of everything

Where will all that expansion end? If the universe has a beginning, it's natural to ask whether it will have an end. Some mythical cosmologies produce endings that are apocalyptic, or recycle the universe to make another; again, scientific cosmologies follow suit.

Big Bang, Big Crunch

The oscillating universe sees the universe expanding and then contracting. This was favoured by Einstein after Hubble's discovery made his original model untenable. An updated version of the model was published in 2002 by Paul Steinhardt and Neil Turok and called the ekpyrotic model after the Stoics' theory of a universe intermittently engulfed in fire (from the Greek *ekpurōsis*, conflagration). It proposes a sequence of 'bounces': Big Bang – expansion – contraction – Big Crunch – Big Bang. This sequence is repeated infinitely. It avoids the tricky 'what was there before?' question, as there is no 'before' in a perpetual cycle.

Given that we know the universe to be expanding, it can either alternate between expanding and contracting or it can keep expanding. If it keeps expanding, its expansion could eventually slow down and stop, or it could continue until the universe is a vast, cold desert of widely separated particles. Eventually, even the smallest particles will be torn apart. Another possibility, labelled the Big Rip, is that expansion will speed up until eventually it rips the universe apart. The most recent prediction for this possibility, published by Chinese astronomers Zhang Xin and Li Miao in 2012, is that it could happen as soon as 16.7 billion years from now (though 103 billion years is more likely).

Other worlds, other universes

Aristotle believed the world to be unique – it was impossible for there to be another. All the four elements existed only in the sub-lunar sphere (the area within the Moon's orbit, including Earth). The Bible, too, seems to suggest that Earth is a unique creation. Yet the possibility of other worlds is very ancient. When in 1277 the Catholic Church issued its list of Condemnations – ideas that were banned – one of the opinions that could lead to excommunication was that God could *not* make other worlds. It was accepted that he probably had not done so, but to suggest that he could not do so if he wished was not allowed.

The Chinese philosopher Deng Mu (1247–1306) proposed not just other worlds, but other heavens. In the 20th century, some scientists proposed other universes. The multiverse – the collection of postulated alternative universes – is not even definitely a theory, never mind definitely a thing. While some respected astrophysicists, including Stephen Hawking, support the theory, others

The Big Bang could either lead to continuing, eternal expansion, or could reverse after a certain point and result in a 'Big Crunch'.

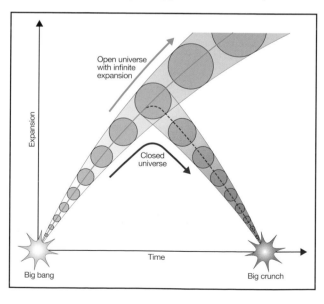

Open universe with infinite expansion

Closed universe

Expansion

Time

Big bang

Big crunch

An artist's concept of the Giant Magellan Telescope, which will be situated in the Atacama Desert some 115km (71m) north of La Serena, Chile. Super-telescopes such as this should provide more information on the birth of the universe and possible life elsewhere.

claim it is a philosophical idea rather than a theory. A scientific theory must be falsifiable (capable of being proven wrong by experiment or observation), yet it is hard to see how the multiverse could be proved not to exist.

> 'Heaven and earth are large, yet in the whole of empty space they are but as a small grain of rice. . . . How unreasonable it would be to suppose that besides heaven and earth which we can see there are no other heavens and no other earths?'
>
> Deng Mu, *The Lute of Bo Ya*, 13th century

In 1952, German quantum physicist Erwin Schrödinger (1887–1961) said that the equations he had formulated didn't show possible alternatives for one history, but that all the histories happened simultaneously. This is the first inkling of multiverse theory. It was formulated by American physicist Hugh Everett III (1930–82) in 1957 and is commonly known as the 'many worlds' interpretation of quantum theory. In essence, it suggests that all possible histories are true; so everything that could ever have happened, has happened, in an infinite series of alternative universes. There are now, appropriately, many versions of multiverse theory. But they more properly belong to quantum theory than to astronomy.

CHAPTER 8

The final
FRONTIER

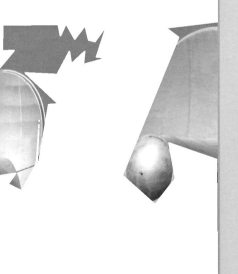

'*We must therefore admit again and again that elsewhere there are other gatherings of matter such as this . . . [and] in other parts of the universe there are other worlds and different races of men and species of wild beast.*'

Lucretius,
On Nature, 1st century BC

As we have been displaced from our treasured position at the centre of the universe and of creation, it is perhaps natural that we begin to wonder what or who else might share the universe with us. One of the newest branches of astronomy is astrobiology – the study of how life evolved and its possible existence elsewhere in the universe.

The Allen Telescope Array, funded by Paul Allen, co-founder of Microsoft, has a large number of small dishes, specifically designed to help in the search for signals indicating extra-terrestrial intelligence.

Asking 'Is there anybody there?'

The idea that there might be other beings elsewhere in space, whether in our solar system or further afield, is far from new. Lucretius wrote of the possibility in the 1st century BC. Giordano Bruno proposed, among other incendiary heresies, that the universe is infinite and contains an infinite number of other worlds that are home to other intelligent beings. It's not clear whether it was this that provoked the Church to burn him for heresy in 1600, since the file relating to his trial and execution is missing from the archives, but it probably didn't help.

Some 86 years after Bruno was executed, Bernard le Bovier de Fontenelle published *Entretiens sur la pluralité des mondes* ('Conversations on the plurality of worlds').

> *'The universe has 10 million, million, million suns (10 followed by 18 zeros) similar to our own. One in a million has planets around it. Only one in a million million has the right combination of chemicals, temperature, water, days and nights to support planetary life as we know it. This calculation arrives at the estimated figure of 100 million worlds where life has been forged by evolution.'*
>
> Howard Shapley, Professor of Astronomy, Harvard University, 1959

It was written in French and translated into English the following year; the author had none of the problems that Bruno had attracted just a century earlier. Thereafter, several authors of scientific texts and fiction considered the possibility of extraterrestrial life. When Giovanni Schiaparelli claimed to have discovered *canali* on Mars, the public soon began to speculate about the possibility of life on the Red Planet on the grounds that if aliens can build canals they must have a technology and therefore be intelligent. Percival Lowell dedicated a personal fortune to investigating Mars (see page 135).

Enthusiastic advocates for the search for alien intelligence have included the American astronomers Carl Sagan (1934–96) and Frank Drake (b.1930). Drake was also responsible for framing the Drake Equation – a formula for calculating the

Giordano Bruno held many heretical beliefs, among them that there could be an infinite number of inhabited worlds besides our own.

THE DRAKE EQUATION

The Drake equation is:

$$N = R_* \times f_p \times n_e \times f_l \times f_i \times f_c \times L$$

| N – the number of civilizations in our galaxy with which communication might be possible | R$_*$ – the average rate of star formation per year in our galaxy | f$_p$ – the fraction of those stars that have planets | n$_e$ – the average number of planets that might support life, per star that has planets | f$_l$ – the fraction of these planets that actually go on to develop life at some point | f$_i$ – the fraction of these planets that go on to develop intelligent life | f$_c$ – the fraction of civilizations that develop a technology which releases detectable signs of their existence into space | L – the length of time for which such civilizations release detectable signals into space |

Very few of the variables could be quantified when Drake wrote the equation in 1961. More recent data suggests that the rate of star formation in the Milky Way is about seven per year, that most stars have planets (so the fraction is close to 1) and that about a fifth of stars have planets in the habitable zone – though that does not in itself mean they can support life. The other variables remain unquantifiable.

likelihood of radio communication with aliens (see box, above). Although more values can be filled in now than when Drake formulated his equation, there are still many unknowns. Depending on the values used for the unknown variables, the number of intelligent alien civilizations currently capable of interstellar communication comes out at anywhere between near zero and many millions.

'Where is everybody?'

The first proposal for contacting alien life was made by Friedrich Gauss in 1802. He suggested that humans could signal to beings on Mars by drawing giant symbols in the snow of Siberia. The idea was not put into practice. During the 20th century, though,

finding or communicating with alien life started to look feasible. It began to be treated seriously after the mathematician and physicist Enrico Fermi asked an innocent question over lunch in 1950: 'Where is everybody?' Given that there are so many stars, probably many with their own planets, why have we not yet encountered aliens? This has become known as the Fermi Paradox.

Fermi's question prompted the Search for Extra-Terrestrial Intelligence (SETI), formally established as a program by NASA in 1971. The SETI Institute was spun off as a separate, not-for-profit organization in 1984. The object of the SETI project is to use radio telescopes to scan the sky for signals from outer space and look for anything that might be a deliberate communication,

THE WOW! SIGNAL

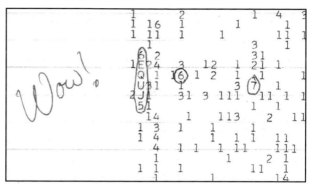

In 1977, SETI researcher Jerry Ehman was studying data from the Big Ear Radio Telescope at Ohio State University. He found a sequence that showed a signal at the frequency of hydrogen emissions growing suddenly stronger and then fading away again. The signal lasted 72 seconds. Ehman circled it on the print-out, writing 'Wow!' beside it in red. If the signal was sent by aliens, it makes sense that it was on the same frequency as hydrogen – it's a part of the spectrum where any civilization searching for contact would look. The signal has never been seen again.

There is no compelling explanation of it even now. Some astronomers still believe the Wow! signal could have come from an extraterrestrial civilization. Others say that a repeated signal would be expected if that were the case – but we Earthlings never repeated the Arecibo message, so that's not inevitably true.

or evidence of a signal broadcast by an intelligent civilization. So far, only one possible signal has been detected (see box, above). Although it is more likely that there is microbial alien life, at least in abundance, this will be harder to find at a distance than a civilization that can use electromagnetic radiation to communicate.

Listening and talking (to aliens)

Communication is a two-way process. We have also sent messages into space for potential alien eyes and ears, but with far less rigour and method. The first was the Arecibo message sent in 1974 from the transmitter at the Arecibo Radio Telescope in Puerto Rico. It comprised a simple pictorial message of 73 lines, each with 23 characters (1/0). As 73 and 23 are both prime numbers (divisible by one and themselves only), this should alert anyone detecting the signal that it's deliberate and worth looking at. It was broadcast directly towards the M13 star cluster, containing around a third of a million stars. However, M13 is on the edge of the Milky Way, 21,000 light years away, so we would need to wait 42,000 years for a response.

Other messages have been targeted at much closer stars, between 17 and 69 light years away. The first to reach its destination will be the RuBisCo Stars message, due to arrive at Teegarden's star in 2021. It contains the DNA sequence of the protein RuBisCo which is responsible for photosynthesis in green plants. Photosynthesis is the ultimate source of energy for all life on Earth.

Not all messages sent to possible alien cultures have been so intelligent. They have included a bunch of Twitter messages, a Beatles song and an advert for Doritos.

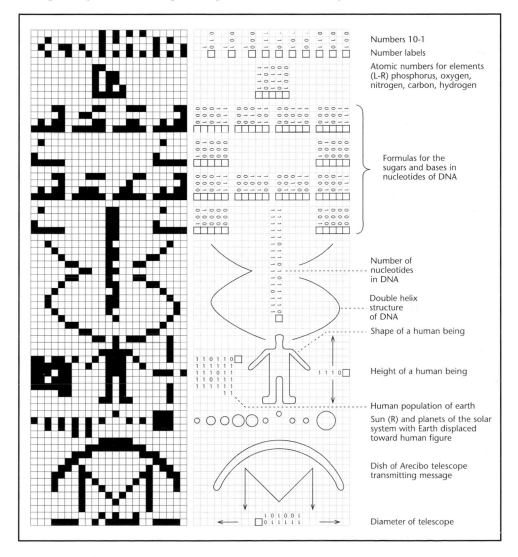

The Arecibo message includes the atomic numbers of crucial elements, molecular information about DNA, and crude images of a human form and the telescope that sent the message.

CONTACT?

The movie *Contact* (1997), based on a 1985 novel by Carl Sagan, is based on the premise that the first TV broadcast with a signal strong enough to leak into space was picked up by aliens on a planet orbiting the star Vega, located 26 light years from Earth. It was the opening of the Berlin Olympic Games in 1936.

Earth is leaking electromagnetic radiation all the time and has been doing so since the first radio and television broadcasts that had sufficient strength to get beyond Earth's ionosphere. But these are not directed at any target so, like the ripples from a stone thrown into a pond, they grow ever weaker as they spread out in all directions. Newer digital transmissions have lower energy and will have even shorter interstellar reach. There's not much chance of an alien near Vega setting out to visit the Berlin Olympics.

Into the void

While it's relatively easy for us to broadcast a radio signal into space, it's impossible to tell whether an alien intelligence would understand the message. To facilitate this understanding, there is an entire field of research devoted to making a code as easy as possible to break – the opposite of encryption. One way is to send a signal that acts as an alert or welcome banner before the message itself.

Another method is to send a physical message into space. The object itself acts as the banner, but the message still needs to be understandable (using pictures may help). But it's akin to launching a message in a bottle, as it will only be found if an alien intelligence encounters the spacecraft carrying the message. The first such project was a plaque attached to the space probes Pioneers 10 and 11, launched in 1972 and 1973 respectively. The second was the Golden Record that accompanied the two Voyager spacecraft in 1977. All are heading out into space beyond the solar system and will continue to do so unless they are destroyed.

Each Pioneer probe bears a plaque on the outside designed for alien eyes. It shows nude male and female human figures; the configuration of a hydrogen molecule; the position of the Earth in the solar system and the position of the Sun relative to 14 pulsars and the centre of the Milky Way. Contact with Pioneer 10 was lost in 2003 at a distance of 80 AU (132,000km/82,021 miles) from Earth.

The two Voyagers are still in contact with Earth; it currently takes 17 hours for signals from them to arrive.

Both Voyagers carry a copy of the Golden Record, and an instrument to play it. The disk holds 115 images of Earth, recordings of natural sounds including wind, thunder, surf, bird song and whale song; music from different times and places; and greetings spoken in 55 languages including Akkadian, which was last spoken 4,000 years ago. Some of the messages are rather endearing:

'Hope everyone's well. We are thinking about you all. Please come here to visit when you have time.' (Mandarin Chinese)

'Hello to everyone. We are happy here and you be happy there.' (Rajasthani)

'Dear Turkish-speaking friends, may the honours of the morning be upon your heads.' (Who knew aliens spoke Turkish?)

But some are downright alarming:

'Friends of space, how are you all? Have you eaten yet? Come visit us if you have time.' (Amoy, Min dialect)

The record also includes a diagram showing the location of our Sun, which will help any hungry extraterrestrials find the dinner table. The Golden Record comes with a stylus and instructions on how to play it that any advanced civilization should be able to decode.

A spot of uranium-238, which has a half-life of 4.51 billion years, will help any aliens work out when Voyager was made; measuring how much is left undecayed

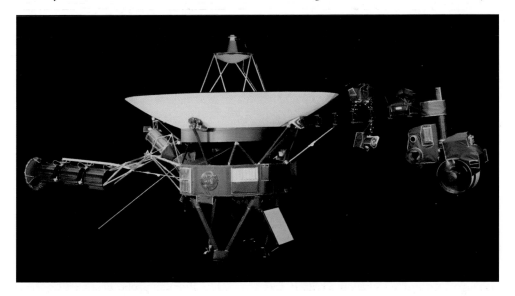

The Golden Record is visible on the outside of Voyager. The outer side will be 98 per cent intact after 10,000 years and the reverse should last a billion years.

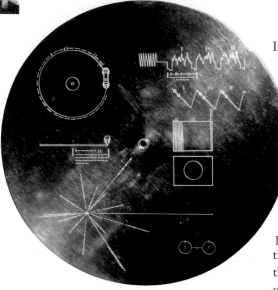

The outside of the Golden Record includes instructions for aliens on how to play the record and what the first screen should look like (a circle on a rectangular screen). The star shape bottom left shows the position of the Sun relative to 14 pulsars.

will tell them how long ago the craft was launched (unless it's all gone, of course).

Voyager 1 is now the most remote human-made object, already more than 124 AU (18 billion km) from Earth. It is heading out of the solar system, into interstellar space, at the rate of 61,350 kph (17km per second).

If Voyager 1 were heading for the nearest star (which it's not), it would take 73,775 years to get there. The next star it will approach is AC+79 3888, but its closest pass will be 1.6 light years (15 trillion km). Voyager 2 will pass within 4.3 light years of Sirius, the brightest star in the sky – but not for 296,000 years.

Some noted scientists have publicly doubted the wisdom of sending messages into space advertising our presence. Stephen Hawking has suggested that it could draw the attention of aliens that might attack us, asset-strip our planet or otherwise act in an unfriendly way. And yet he acknowledges that we won't stop looking and hoping.

Somewhere to live

But where to look? If aliens are out there, they live somewhere, probably on planets that are somewhat similar to ours, rather than, say, gas giants. The way to look for alien life that is not sufficiently advanced to send radio signals is to find the planets it might inhabit. The search for exoplanets – planets outside our solar system – began in earnest in the 1990s.

'Look again at that dot. That's here. That's home. That's us. On it everyone you love, everyone you know, everyone you ever heard of, every human being who ever was, lived out their lives. The aggregate of our joy and suffering, thousands of confident religions, ideologies, and economic doctrines, every hunter and forager, every hero and coward, every creator and destroyer of civilization, every king and peasant, every young couple in love, every mother and father, hopeful child, inventor and explorer, every teacher of morals, every corrupt politician, every "superstar", every "supreme leader", every saint and sinner in the history of our species lived there – on a mote of dust suspended in a sunbeam.'

Carl Sagan, 1994

HOW TO FIND AN EXOPLANET

There are several ways to spot an exoplanet, but they are too small, dim and distant to see directly with even a super-powerful telescope. Instead, astronomers look for tell-tale signs that a star has planets. There are several ways a planet can give itself away.

- When a planet passes in front of its star, it blocks the light from the star briefly, making the star look dimmer for the duration of the transit.
- Planets can make a star 'wobble' slightly. As the planet's gravity affects the star, so the entire star-planet system moves from side to side as the planet orbits the star.
- Planets also act like a brake on their star, slowing its rotation. If a star of known size and composition spins more slowly than expected, planets are suspected.
- The youngest known exoplanet, only a million years old, orbits a star called Coku Tau 4. It was revealed by a gap in the disc of dust surrounding the star; the gap is made by the planet drawing matter towards itself by gravity as it forms.

Known as the 'pale blue dot', this photo was taken by Voyager at Carl Sagan's request when the craft was 6 billion km (3.7 billion miles) from Earth.

Those infinite worlds

The first exoplanet was discovered in 1992, orbiting the pulsar PSR B1257+12. A pulsar is the remnant of a star that has reached the end of its life (see page 164). Such a planet is relatively unlikely to host life. The first exoplanet orbiting a main-sequence star was confirmed in 1995. Called 51 Pegasi b, it orbits its star every four days and is just over 50 light years from the solar system. Since then, around 3,500 exoplanets have been found and confirmed, with thousands more marked for investigation but not yet confirmed. Most of them have been found using the Kepler space telescope launched in 2009. Kepler constantly monitors and photographs 145,000 main-sequence stars in a fixed field of view.

PSR B1257+12 A, B and C are three planets orbiting the pulsar B1257+12. They were renamed Draugr, Poltergeist and Phobetor in 2015. The pulsar rotates once every 6.22 milliseconds (9,650 rpm).

The Goldilocks zone and other factors

Aliens aren't going to live just anywhere. There are certain conditions that astrobiologists think life probably needs, such as temperatures within a certain range and the presence of liquid water. This rules out many of the hundreds of billions of planets that probably exist in the universe. The idea that there is a zone around some stars where planets capable of supporting life might be found emerged in 1953 with the work of Harlow Shapley and, independently, the German physiologist Hubertus Strughold (1898–1986). Both stressed the necessity of water to life. The Chinese-American astrophysicist Su-Shu Huang (1915–77) introduced the term 'habitable zone' in 1959 and studied the type of stars and planets that could provide habitable environments. The concept was further developed during the 1960s, with the term 'Goldilocks zone' emerging in the 1970s to describe the zone around a star in which planets that are 'just right' in terms of temperature can be found. In 2013, a model for a circumplanetary habitable zone was proposed to include the area in which moons might be able to sustain life. In 2000, the American palaeontologist Peter Ward (b.1949) and astronomer David Brownlee (b.1943) extended the habitable zone concept to galaxies, defining an area that is neither too central nor too far out from the centre of a galaxy as being the area where stars might have life-supporting planets. Their 'rare Earth' thesis argues that in fact advanced life is very rare in the universe, and possibly even unique to Earth.

The next big thing

Of course, extraterrestrial life might dwell inside the solar system, too. NASA is now

'We believe that life arose spontaneously on Earth, so in an infinite universe, there must be other occurrences of life. Somewhere in the cosmos, perhaps intelligent life might be watching these lights of ours, aware of what they mean. Or do our lights wander a lifeless cosmos, unseen beacons announcing that, here on one rock, the universe discovered its existence? Either way, there is no better question. . . . We are alive. We are intelligent. We must know.'

Stephen Hawking, 2015

prioritizing the search for evidence of life in its Mars missions. Some of the moons of the gas giants, such as Saturn's satellite Enceladus with its vast sub-surface ocean, are also possible candidates for hosting simple life. Finding even the simplest of life forms on another planet or moon, in the solar system or elsewhere, would force us to re-evaluate our position in the universe. It would have massive social, psychological and spiritual as well as scientific impact, and be at least as big a paradigm shift as the discovery that there are other worlds, other solar systems, other galaxies.

An Earth-like planet, located in a star's habitable zone, would be neither too hot nor too cold for liquid water. It is in places like this that we should look for alien life.

Index

Picture credits

We have made every effort to contact the copyright holders of the images used in this book. Any omissions will be rectified in future editions.

Bridgeman Images: 22 (Astronomical tablet from Kish, recording the rising and settings of Venus from the first 6 years of the reign of the King of Babylon, 7th century (carved clay), Babylonian/Ashmolean Museum, University of Oxford); 67 (Trial of Galileo, 1633 (oil on canvas) (detail of 2344), Italian School, (17th century)/Private Collection); 116 (Apollo – Helios driving the chariot of the Sun, 1517–1518 (fresco), Peruzzi, Baldassarre (1481–1536)/Villa Farnesina, Rome, Italy/Ghigo Roli); 140 (Leonid Meteor Shower of 1833, USA, Illustration/J. T. Vintage)

European Southern Observatory: 157 (G. Huedepohl/atacamaphoto.com); 161; 172–3 (NASA/ESA/Josh Lake)

Getty Images: 7 (Science & Society Picture Library); 8–9 (Ismail Duru/Anadolu Agency); 16, bottom (Schellhorn/ullstein bild); 26 (DEA/Archivio J. Lange); 28 (DEA/G. Dagli Orti); 31 (DEA/G. Dagli Orti); 35 (Werner Forman/Universal Images Group); 38 (Werner Forman/Universal Images Group); 39 (Bettmann); 41 (SeM/UIG); 43 (top); 58 (DEA/G. Dagli Orti); 66 ((Science & Society Picture Library); 74 (Science & Society Picture Library); 75 (Leemage/Corbis); 77 (Universal History Archive); 88 (J. W. Draper); 103 (Alessandro Vannini/Corbis); 117 (Stefano Bianchetti/Corbis); 120 (De Agostini Picture Library); 124 (DEA/V. Pirozzi); 135 (Ann Ronan Pictures/Print Collector); 139 (Oxford Science Archive/Print Collector); 141 (Elizaveta Becker/ullstein bild); 155, top (De Agostini Picture Library); 175 (Universal Images Group); 178 (DEA/G. Dagli Orti); 187 (Jay M. Pasachoff); 199 (NASA/Hulton Archive)

Google Earth: 16 (top)

Institute of Astronomy, University of Cambridge: 85; 113

P. Frankenstein/H. Zwietasch; Landesmuseum Württemberg, Stuttgart: 13 (top)

Mary Evans Picture Library: 93, 148

NASA: 106, 148, bottom (JPL); 166 (R. Kirshner/Harvard-Smithsonian Center for Astrophysics); 171; 182 (N. Benitez, JHU; T. Broadhurst, Racah Institute of Physics/The Hebrew University; H. Ford, JHU; M. Clampin, STScI; G. Hartig, STScI; G. Illingworth, UCO/Lick Observatory; the ACS Science Team; and ESA); 185 (ESA/M. Kornmesser)

National Archaeological Museum, Athens: 71 (bottom)

Science Photo Library: 25 (Humanities and Social Sciences Library/Asian and Middle Eastern Division); 36–7 (NYPL/Science Source); 63 (Humanities and Social Sciences Library/Rare Books Division/New York Public Library); 72 (Jose Antonio Penas); 87 (Detlev van Ravenswaay); 95 (bottom); 104 (bottom); 125; 127, top (Royal Astronomical Society); 138 (Mark Garlick); 144 (Royal Astronomical Society); 145 (Royal Astronomical Society); 149 (ESA/Rosetta/Philae/CIVA); 152 (Library of Congress); 154 (New York Public Library); 156 (Science Source); 160 (Emilio Segre Visual Archives/American Institute of Physics); 165 (bottom); 188 (Mark Garlick); 197; 201 (NASA)

Shutterstock: 6; 10; 12 (bottom); 13 (bottom); 15x2; 17; 19; 29; 33; 40; 47 (top); 53; 54; 57; 68–9; 70; 73; 82; 91; 96–7; 99; 101; 102; 111; 118; 122–3; 126; 129; 130; 132; 133; 134; 137 (bottom); 147; 155 (bottom); 164; 165 (top); 168; 170; 174; 177; 181; 192–3; 198

Wellcome Trust: 18 (top); 34; 47 (bottom); 61; 159 (bottom)

Page 56: Copernicus, Nicolaus, 1473–1543. 'De revolutionibus orbium coelestium, solar system.' (1566) Rice University: http://hdl.handle.net/1911/78815

Page 176: photograph © Andrew Dunn

Diagrams by David Woodroffe: 11; 12 (top); 43 (bottom); 45; 64; 71 (top); 83; 100; 107; 109; 114; 115 (top); 180; 186; 190; 195